Margaret Plues

Rambles in Search of Flowerless Plants

Margaret Plues

Rambles in Search of Flowerless Plants

ISBN/EAN: 9783337034931

Printed in Europe, USA, Canada, Australia, Japan

Cover: Foto ©berggeist007 / pixelio.de

More available books at **www.hansebooks.com**

RAMBLES

IN SEARCH OF

FLOWERLESS PLANTS

BY

MARGARET PLUES

AUTHOR OF "RAMBLES IN SEARCH OF WILD FLOWERS,"
"GEOLOGY FOR THE MILLION," ETC.

THIRD EDITION

LONDON
HOULSTON AND WRIGHT
65, PATERNOSTER ROW
MDCCCLXVIII.

TO

ALL TRUE LOVERS OF NATURE,

THESE SIMPLE RESEARCHES

INTO THE LOWEST FORMS OF VEGETABLE LIFE

ARE AFFECTIONATELY DEDICATED BY

THEIR OBEDIENT SERVANT,

THE AUTHOR.

TABLE OF CONTENTS.

CHAPTER I.

Classification of Plants. Ferns. Polypodiaceæ. Polypodium. Ceterach. Gymnogramma, ... 1

CHAPTER II.

Fossil Ferns. Aspidiaceæ. Polystichum. Woodsia, . 8

CHAPTER III.

Aspidiaceæ continued. Lastræa, ... 15

CHAPTER IV.

Aspidiaceæ continued. Asplenium, . . . 21

CHAPTER V.

Aspidiaceæ continued. Cystopteris. Athyrium. Scolopendrium, 30

CHAPTER VI.

Aspidiaceæ continued. Pteris. Blechnum. Adiantum. Hymenophyllaceæ. Hymenophyllum. Trichomanes. Osmundaceæ. Ophioglossaceæ, ... 37

CHAPTER VII.

Fern Allies. Equisetum. Isoetes. Lycopodium, . . 47

CHAPTER VIII.

Moss Order. Fructification. Summit- and Side-fruited Mosses. Andræa Group. Sphagnaceæ. Bog Moss. Uses of Bog Moss. Phascæ Group. Weissia Group. Gymnostomum. Solomon's Hyssop. Bristle Mosses. Fork Mosses, . . . 57

CHAPTER IX.

Vitality of Mosses. Pottia. Trichostomæ Group. Screw Mosses. Extinguisher Mosses. Hedwigia. Grimmea. Fringe Mosses, 69

CHAPTER X.

Bristle Mosses. Tetraphis. Buxbaum's Moss. Hair-moss Group. Capituli of Androgynum. Thread Mosses. Bryum, . 80

CHAPTER XI.

Thyme Thread-Mosses. Cord Mosses. Bladder Mosses. Apple Mosses. Collar Mosses. Cavern Moss. Flat Fork Mosses. Fissidens. Mungo Park, ... 92

CHAPTER XII.

Side-fruited Mosses. Leucodon. Climacium. Leskea. Omalia. Neckera, ... 103

CHAPTER XIII.

Hypnum Group. Feather Mosses, . 109

CHAPTER XIV.

Moss Allies. Hepaticæ Order. Jungermanniæ. Foliaceous and Frondose Liverworts. Stipules. Marchantia. Riccia. Anthoceros. Targonir. Nardoo, ... 118

CHAPTER XV.

Seaweed Classification. Olive, Red, and Green Weeds. Melanosperms. Gulph Weed. Cystoseira. Fucus. Kelp. Sea Thong. Laminariæ Order. Desmarestia, ... 127

CHAPTER XVI.

Dictyotaceæ Order. Peacock Weed. Chordariaceæ Order. Mesogloia. Myrionema. Cladostephus. Ectocarpaceæ Order. Sphacelaria, ... 137

CHAPTER XVII.

Red Weeds. Rhodosperms. Odonthalia. Rytiphlæa. Polysiphonia. Laurenciaceæ Order. Chylocladia. Jania. Coralline. Nullipores. Delesseria. Nitophyllum. Ploeamium, . . 145

CHAPTER XVIII.

Dulse. Rhodymeniaceæ Order. Sphærococcus. Gracelaria. Hypnea. Cryptonemiaceæ Order. Gelidium. Edible nests of Japan. Gigartina. Carrageen Moss. Gymnogongrus. Furcellaria. Dumontia. Kallymenia. Cruoria. Ceramiaceæ Order. Ptilota, 154

CONTENTS. vii

CHAPTER XIX.

Green Weeds. Chlorosperms. Codium. Bryopsis. Vaucheria. Confervaceæ Order. Cladophora. Conferva. Ulvaceæ Order. Enteromorpha. Ulva. Bangia. Rivularia. Calothrix. Specimens of Melanosperms, Rhodosperms, and Chlorosperms, . 165

CHAPTER XX.

Freshwater Algæ. Land and Freshwater Vaucherias. Botrydium. Chætophora. Draparnaldia. Chroolepus. Freshwater Confervæ and Ulvaceæ. Batrachospermum. Nostoc. Palmella. Red Snow. Chara. Uses of Freshwater Chlorosperms. . . 174

CHAPTER XXI.

Lichens. Gonidia. Crustaceous and Frondose Lichens. Uses of Lichens. Crustaceous Lichens. Apothecia. Bæomyces. Goblet Lichens. Arthonia. Writing Lichens. Verrucaria. Endocarpon. Pertusaria. Thelotrema. Variolaria. Urceolaria. Ruskin on Lichens, ... 187

CHAPTER XXII.

Leprariæ. Lecidea. Leanora. Cudbear. Scurf Lichens. Squamariæ. Parmelia. Crotal, ... 199

CHAPTER XXIII.

Jelly Lichens. Collema. Dog Lichen. Peltidea. Socket Lichens. Gyrophoræ. Tripe de Roche. Iceland Moss. Orchil. Borrera, 212

CHAPTER XXIV.

Evernia. Ramalina. Usnea. Rock Hair. Coral Lichens. Globe Lichens. Rein-deer Moss. Cup Moss. Leighton's Classification. Lichina, ... 252

CHAPTER XXV.

Fungi. Prejudices. Parts. Classification. Agaric Group. Fairy ring mushroom. Edible mushroom. Chantarelle, . . 239

CHAPTER XXVI.

Polyporei group. Sap balls. Polypores. Trametes. Dædalia. Merulius. Dry rot. Fistulina. Hydnum. Auriculini group. Thelephora. Craterellus. Corticium. Clavariæ group. Calocera. Tremella group. Exidia. Jew's Ear. Dacrymyces, . 260

CHAPTER XXVII.

Gasteromycetes or Envelope class. Hypogei group. Bath Truffle. Phallus. Clathrus. Trichogastre group. Earth stars. Puff-balls. Uses and Legends. Myxogastre group. Arcyria. Trichia. Nidularia group. Sillercups. Thelebolus, . . . 273

CHAPTER XXVIII.

Coniomycetes or Dust class. Epiphytes. Puccinea. Uredo. Smut. Bunt. Yeast plant. Æcidium. Hyphomycetes or Thread class. Moulds, ... 282

CHAPTER XXIX.

Ascomycetes or Ascus class. Elvellacai group. Morel. Helvella. Mitrula. Spathularia. Leotia. Geoglossum. Peziza. Bulgaria. Truffles. Ergot. Xylaria. Hypoxilon. Sphæria, . . 292

Plate 1.

1 Common Polypody 2 Oak Polypody 3 Beech Polypody 4 Limestone Polypody 5 Scaly Spleenwort
6 Gymnogramma 7 Angular Prickly Shield Fern 8 Parsley Fern

RAMBLES

IN

Search of Flowerless Plants.

CHAPTER I.

FERNS.

"All about in shady places the Ferns were busy untucking themselves from their graveclothes, unrolling their mysterious coils of life, adding continually to the hidden growth as they unfolded the visible. In this they were like the other revelations of God the Infinite."—DAVID ELGINBROD.

"On every side spring Ferns, whose feathery leaves
Seem wafted by the perpetual breath of God."

HAVING carefully studied flowering plants, and collected specimens from all the orders of the Two-lobed (Dicotyledonous) and One-lobed (Monocotyledonous) classes, we are now free to enter on the study of the third class of plants, the FLOWERLESS or LOBELESS (Acotyledonous) class. Here we lose the grand distinguishing feature, the flowers, and must direct double attention to the seed, now very minute, and termed *spores*. As the Two-lobed class are further characterised by the outward growth of the main stem or trunk, and are therefore called Exogens, and the One-lobed class by the inward growth, expressed in the term Endogens, so the Lobeless have their distinguishing feature in this respect, adding to their growth by additions to the summit, and thus called *Acrogens* or Summit-growers.

The first natural order in the Lobeless class is that of the Ferns (Filices.) In this order we still find woody fibre, though the great bulk of the plants of the class are formed entirely of cellular tissue. But the Ferns are highly organised, approaching Monocotyledons in some of their features, and having their leaves or fronds beautifully veined.

The parts of a Fern are—1st, the *root*, which we can easily recognise, being subject only to the variations which we observe in the roots of other plants. It has a *rhizome*, which generally creeps upon or under the ground, and might easily be mistaken for the root; and then in tree Ferns there is the bole or stem, which is called a *stipe*. The leaf of the Fern is called a *frond;* and the frond is composed of a main stalk (*rachis*), and leaflets (*pinnæ*.) It has no flowers, and its seeds or spores grow abundantly on the back of the frond. The spores are enclosed in *cases*, which form masses called *sori*, or spore-masses. From the form and position and covering of these spore-masses the genera of Ferns are decided. The first division of Ferns, the Polypodiaceæ, have the seed-cases in round masses, without any covering.

POLYPODIUM.

Our first acquaintance with this group was made in one of the beautiful "dales" of Yorkshire. The river Swale winds serpent-like along the valley, and when we began our exploration the morning sun was turning its waters to gold. On the hill-sides on either hand are deep clefts, worn by mountain streams, the steep banks covered with Birch-wood. In these wooded glades we began our eager search for Ferns. For some time each Fern that we gathered seemed too complex in its structure for our zeal to cut its teeth upon; but presently we recognised the common Polypody (Polypodium vulgare, *Plate I.*, *fig.* 1), and some of our party seized it, exclaiming, "The seeds here are of a sensible size; one can discern the little clusters of grains without a glass." We were, however, determined at once to accustom ourselves to the use of the pocket-lens, and by its aid we saw that Fanny's "grains" were cases of spores. The spores themselves looked like fine dust. The entire absence of covering (indusium), proved the right of the Fern to stand in the Polypody group. This is one of the commonest Ferns that grow. It is found on walls or old stumps; its branching rhizome matted with moss, and its fronds assuming every graceful bend and curve.

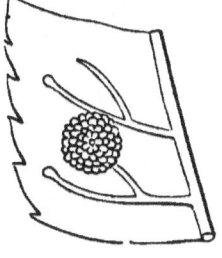

POLYPODIUM.

There is a Welsh variety of this plant (P. cambricum),

broader in growth, and with the pinnæ cleft, but it is never found with spores.

The Irish variety (P. hibernicum), is doubly-divided (pinnatifid), and fertile.

The common Polypody is not very good for garden ferneries, because its fronds perish in the first frost, according to Sowerby. Under shelter it is evergreen. A medicine made from it used to be given in whooping-cough.

Further on in the dell, where the trees made a deep shade, a quantity of a beautiful small Fern was growing like a miniature forest. Its foliage varied from dark green to the tender tint of the first spring foliage. The rachis was slender and brittle, from half a foot to a foot high, and terminating in three branches. Here and there one frond stood higher than the rest, with its pinnæ much curled-in. On examining some of these we found them plentifully speckled with spore-clusters, while the more fragile undergrowth were destitute of fruit. Our lens shewed these to be coverless, so we concluded that it must be a Polypody; and its three branches and rachis, coloured with purplish brown, indicated it to be the Oak-Fern (P. dryopteris, *Plate I., fig.* 2). The only objection to this conclusion was that it was growing in a Birch-wood, and the Oak-Fern is generally described as inhabiting Oak-woods. We were relieved from this difficulty by finding one or two small Oaks in the vicinity; but the leaf mould in which its rhizome was lightly rooted, branching in every direction like the underground stem of the Wood Anemone, was chiefly compounded of Birch leaves. This fern is one of the most easily cultivated. I have seen it growing upon the ledges of rockeries like plantations of young Firs, spreading freely, and maintaining the easy grace of its native growth.

On a spongy bank, where the soil was more clay-like, was another miniature tree Fern. The size was somewhat larger than that of the Oak-Fern, and the rachis more robust. The frond was triangular in form, the upper pinnæ being short, and lengthening gradually, the last two bending forward. The naked groups of spore-cases showed this to be also a Polypody, and the forward bend of the lowest pinnæ proved it to be the Beech-Fern, or Polypodium phegopteris (*Plate I., fig.* 3). This is more difficult to cultivate than the Oak Polypody, being more dependent upon shade; but when it gets once established it continues to flourish, though not increasing nearly so quickly as its more domestic brother. We have since found both these Ferns in the Scotch highlands.

In a glen higher up the dale, the Limestone Polypody (P. calcareum, *Plate I., fig.* 4), flourishes. Its three branches resemble those of the Oak-Fern; but it is more rigid in its manner of growth, attains a larger stature, has a green rachis, and a powdery appearance. Fanny has found it in abundance among the combes of Somersetshire. Limestone seems necessary to its comfort. We see no habitats assigned to it where that is not the prevailing rock; indeed, Sowerby asserts that where other rocks top the limestone, the Fern avoids them, though growing abundantly in their near vicinity. This

Fern is easily cultivated, flourishing hardily wherever the soil is well drained.

The Alpine Polypody (P. alpestre), was sent to us by a Shropshire ally. In form it resembles the Lady-Fern, The pinnæ are planted all along the rachis, very short at the summit, long in the centre, and becoming short again at the base; they extend for more than three-fourths of the rachis. The specimen given to us was brought from the Shropshire hills. It was beginning to die away even before the delicate fronds of the Lady-Fern, its neighbour, showed the least sign of decay. No cover was upon the seed-masses then, but its neighbour had also parted with her's. Our friend placed it in the fernery, and watched for next year's fruit; but, alas! before it was fairly developed a greedy cow, trespassing into the garden, selected its fronds as its *bonne bouche*. The third year was more fortunate—the fructification was formed, and no indusium was discernible. So the Shropshire mountaineer takes its place among the Polypodies.

The near ally of this family, the Scaly Spleenwort (Ceterach officinarum, *Plate I., fig. 5*), I found at Vallis in Somersetshire, growing on walls. Afterwards I found it abundantly in Devonshire, and about Congresbury and Yatton, in Somersetshire, generally in company with the Black-stalked Spleenwort and Wall Rue, making the walls into botanic gardens. It is a compact little plant; the fronds thick, and cut into broad simple pinnæ. They are lined with

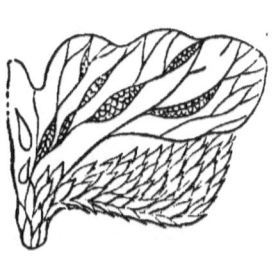

CETERACH.

something like brown felt, being small semi-transparent scales, white in their youth, but presently becoming brown. These scales seem to protect the seed-vessels, and serve them instead of covers.

The Jersey-Fern (Gymnogramme leptophylla, *Plate I.*, *fig.* 6), we none of us have found. We considered ourselves lucky enough in getting a sight of it. It has little oval pinnæ, cut and toothed. The fruitful fronds are more erect than the others, and have narrower pinnæ.

The Parsley-Fern (Allosorus crispus, *Plate I.*, *fig.* 8), claims alliance with the Polypodies. Its spore-masses are uncovered, except for the margins of the pinnæ, which try to overlap the line of clusters. We found a small plant of this in a beautiful table-land called Stag-Fell—supposed to have been the last Yorkshire retreat of the wild deer. It lies between Swaledale and Wensleydale, and the Road that traverses it is called the Butter-tub Pass. It was a mere accident that the Parsley-Fern should be found there, for it generally avoids the limestone, and it was a very poor specimen. The lake country boasts it in great abundance. There it grows in free clusters among the sward, the fronds attaining a height of eight or ten inches. The fruitful fronds grow erect, the barren ones clustering beneath them, very much like the plant from which it takes its name.

ALLOSORUS.

CHAPTER II.

FERNS.

'The feathery Fern! the feathery Fern!
　It groweth wide, and it groweth free,
　By the rippling brook, and the wimpling burn,
　　And the tall and stately forest tree;
　Where the merle and the mavis sweetly sing,
　　And the blue jay makes the woods to ring,
　And the pheasant flies, on whirling wing,
　　Beneath a verdurous canopy."　　　ANNE PRATT.

SOON after our first essay in the study of Ferns, I found an opportunity to steal away quietly into that sweet wood alone. Making my way along a tangled path to a much greater distance than we had penetrated on the former occasion, I passed under some precipitous rocks, and found myself in a shady part of the wood. Here

"Paths there were many,
　Winding through palmy Fern, and Rushes fenny,
　And Ivy banks;"

and I chose one close by the margin of the brook. Huge masses of rock were strewn both in the narrow wood and in the bed of the stream, revealing the fact that, peacefully as its waters now gurgled on, yet that winter storms could make it rush and roar till the whole of the gorge would be converted into a river's bed, and the waters be mighty enough to roll the huge boulders from the hills beyond. The rock on which I seated myself was covered with stony pipes: it was, in fact, a mass of fossil coral: and in another of the boulders I recognised remains of the long-extinct animal lilies: while many had lain so long among the trees there, that they were covered with moss and rock plants, and the graceful Ferns waved proudly over them, like Cypresses, marking the tomb of the corals and encrinites. I was geologist enough to know that another member of the same rock formation which contains these is replete in fossil Ferns and their allies. These early-created plants—these patriarchs of vegetation—form the chief part of the coal measures. Thus we may well regard Ferns as the aristocracy of vegetable life, the "oldest family in the country." To date from William the Conqueror they would consider fungus gentility indeed! Certainly, as we see them in England, they are very reduced and insignificant members of the ancient and honourable house, but they may hold up their heads proudly, and share with the Bruce the motto "*Fuimus.*" In the far back ages, before the coal cellars of the earth were furnished, they were distributed in great quantities over the northern hemisphere, large areas being covered with a multitude of Ferns or Horsetails, of but one or two species; while in other extensive stretches a few other species prevailed. We are told that this is still the case in the southern hemisphere; and in Van Diemen's Land, and New Zealand especially, they grow in such profusion as to choke the young trees, and admit no undergrowth of smaller species, themselves attaining the size of forest trees. In such cases, Orr tells us, the climate is damp and equable, and the variety of Ferns small. Their comeliness of form and lightness of foliage fill the hearts of beholders with adoring wonder—these stately foreign relations uphold the family grandeur in the present age. Thus the Ferns have their family history—legends of obscure light caused by vaporous atmosphere, a grand catastrophe, and a universal tomb; and their past provides light, warmth, and comfort for our present—furnishing coal, gas, and even dyes. Certainly they throw a glowing light upon God's fatherly care in turning the ruins of immature nature into a blessed provision for the creature of His special favour—man.

> " And so the heart, intently gleaning
> O'er fields of legendary lore,
> May light upon a holier meaning—
> A meaning never found before."

I had come into the wood to search for one special family of Ferns, the Polystichums, which come next in order, according to my book, to the Polypodiaceæ. These Prickly Shield-Ferns are of an elongated form; the pinnæ are divided again, or bipinnate, and the masses of seed-cases have round covers, attached by a thread in the centre. One species is decidedly evergreen; the others are so in sheltered situations. They are of a firmer tougher texture than any other of our native Ferns, and should be placed as the vanguard of the fernery, as they bear wind and weather better than any others. Beautiful Ferns were growing in rich profusion around me, but these were triangular in form, and their spore-masses had kidney-shaped covers. Some stones lying in the brook tempted me to cross, and I succeeded in doing so without wetting my feet. Under the deep bank on the other side, the fronds bowing so as nearly to dip into the stream, I espied some Ferns of the long narrow form I was seeking. The plant was about a foot and a half high, eight fronds springing in a circle from the rhizome, each bending outwards, so as to form a basket or crown shape. They were broad in the centre, and tapered to each extremity. In some the pinnæ were placed alternately on the rachis ; in others they were nearly opposite. The leaflets or *pinnules* were scolloped sharply, and a kind of ear at the base of each gave them somewhat of a crescent form. When I gathered a frond I found it very tough, and requiring a good deal of force to detach it. My lens showed me the round covers on the spore-masses, with their central attachment, but I was at first at a loss to know which Polystichum it was. It recurred to my mind having heard a great botanist explain the difference between two of these Ferns. " The angular one," he said, " when held up to the light, showed a clear line between the pinnæ and the stems, while in the acute species the pinnules were so close at the base as to show no light between them." I held one up against the rich light of the setting sun, and the leaflets seemed to run together —another, the same : this, then, was the common Prickly Shield-Fern (Polystichum aculeatum). I found other fronds with the leaflets as finely cut but somewhat broader and more distinctly eared ; holding it up I was gladdened by the sight of the clear line. Here, then, I had a second member of the family—the Angular Prickly

Shield-Fern. (P. angulare, *Plate I., fig.* 7). Very near the brook, further on, I found a Fern with similar characteristics, but the fronds were smaller, the leaflets larger, more eared, and much less numerous, and the colour of a darker and more vivid green. In all these particulars it answered to the description of the Lobed Prickly Shield-Fern (P. lobatum). My book opined that it was a distinct species, and its appearance favours that opinion, though many high authorities consider it merely a variety of the common Polystichum.

POLYSTICHUM.

Returning by the fernery, I ventured to take a frond of the Holly-Fern which Fanny had brought from Llanberis. In this there were no branches. The rachis was set with the large, crescent-shaped, prickly leaflets or pinnules. Its dark glossy hue and prickly edges, together with the fact of its being evergreen, entitle it fully to be named the Holly-Fern (P. lonchitis). It is an Alpine plant, and the only time I ever saw it in its wild state was in a mountain wood near the Gemmi Pass. There the fronds were above a foot long, springing in the coronal form, the clusters looking like verdant baskets among the exquisite variety of Alpine flowers; while snow-capped mountains rose on every side, and a glacier-born torrent rolled down the steep descent, laving the little fern roots as it passed on in its mad career. Here and there a vast assemblage of tree stumps bore testimony to an avalanche having swept over the spot, and carried away the thousand mighty trunks, as the scythe would clear off a curve of grass upon the lawn.

But hardy as the Holly-Fern is upon its native hills, it seldom flourishes for many years in the fernery. This is partly for want of careful drainage; but even where that is attended to, the mountaineer too often dwindles away in its confined position.

The Woodsias come in between the Polypodies and the Polystichums; but we none of us found any specimens then or since. I have seen them in Fern cases, but cannot think of them with anything like the pleasure with which I remember wild specimens.

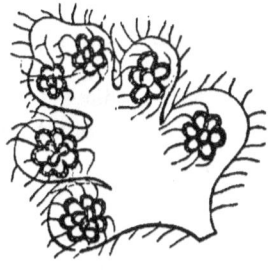

WOODSIA.

The Alpine Woodsia is covered with brownish hairs, especially under the pinnules. The spore-masses are enclosed in a cover, which opens in the centre, and splits into threadlike segments, which surround the seed-clusters like a fringe. I made a careful sketch to place in my collection, until such time as I should succeed in getting a living specimen (*Plate IV., fig. 2*).

The Woodsia ilvensis has been found in Teesdale, a valley not far from the place of my then sojourn—it separates Yorkshire from Durham; but for many years the Fern has been sought there in vain.

Both the Woodsias are rare inhabitants of Alpine situations.

I laid my specimens, large and small, between sheets of botanical paper—a big stone and a packing-case-lid forming a capital extempore press.

We found our new collection a very congenial friend. We loved God for making the beautiful plants. We wondered at His adaptation of them to the service and delight of man. Marvellous that the vegetation of a past age, when man was yet the dust of the earth, should be stored up for his use in the deep bowels of the rocks! Surely each little Fern, reminding us of this bountiful care, teaches us a lesson of reliance upon God, and puts to shame the unbelief of anxious alarmists, as Martin Tupper so plainly points out :—

"Yet man, heedless of a God, counteth up vain reckonings,
  Fearing to be jostled and starved out by the too prolific increase of his kind;
And asketh, in unbelieving dread, for how few years to come
Will the black cellars of the world yield unto him fuel for his winter.
Fear not, son of man, for thyself nor thy seed. With a multitude is plenty;
God's blessing giveth increase, and with it larger than enough."

Plate 2.

1 Male Fern _ 2 Spreading Shield Fern _ 3 Spring Shield Fern _ 4 Heath Shield Fern.
5 Marsh Shield Fern _ 6 Crested Shield Fern.

CHAPTER III.

FERNS.

"The stately Fern, the golden Broom,
　　The Lily tall and fair—
All these in rich succession bloom,
　　And scent the summer air.
In secret dell, by murmuring rill,
　　In gardens bright and gay,
Within the valley, on the hill,
　　They cheer our toilsome way!"

　　　　　　　　　　ILIMON.

TAKING our way across the dell towards the moors, we gathered some tall Ferns by the side of the path. These resembled the Polystichums in form, but grew more erect; still, however, in the basket-style of group. The pinnules were bluntly cut, not sharply like those of the Prickly Shield-Ferns. The pinnæ were placed alternately on the rachis, covering three-fourths of its length; semi-transparent brown scales clustered thickly on the stem where it was free from pinnæ, and thin narrow ones were scattered along it, even where the pinnæ were present. Plentiful spore-masses were sprinkled over the backs of the

LASTREA.

pinnules, ranged in a row on either side between the

midvein and the margin; and the lens showed that the covers on these masses were kidney-shaped. This was the common Male-Fern, the most frequent member of the Lastrea group, characterized by the kidney-shaped cover or *indusium* (L. filixmas, Plate II., *fig.* 1). These Lastreas compose the third family of the second group of Ferns—the *aspidiaceæ*, the Polystichum and Woodsia families preceding it in the group. The English name of the Lastreas is Shield-Fern.

The Spreading Shield-Fern (L. dilatata, Plate II., *fig.* 2), was there also. Growing to the height of two feet, the lower pinnæ becoming so elongated as to give a triangular form to the frond, the pinnules branched again, and beset with independent leaflets, so as to be *tripinnate*, and each little leaflet curled in at the edges, the Fern has an appearance at once stately and graceful. Surely this species must have suggested to the poet the expression, "The palmy Fern's green elegance." The foliage is often beautifully shaded, becoming very pale towards the ends of the little branchlets. I remember being greatly struck with its loveliness on seeing it bending over the margin of Sheerwater Lake, in Wiltshire, where its verdant tint and feathery form were mirrored in the limpid waters. Nor did it show to less advantage beside the yellow Broom on the wooded hills of Herefordshire. Ever as I looked upon them I exclaimed—

"Cool are the Fern-tufts, green their plumes,
    Golden the blossoms on the Brooms."

Indeed, they make the hillside itself golden, as if a cloud of gilded butterflies had settled on the brushwood. Here, in the Yorkshire glen, it looked fresh and graceful, and we were warm in our appreciation of our new acquisition.

By a little gate we passed from the steep wood to the yet steeper pasture ; and, as from time to time we paused to take breath, the view became continually wider and more beautiful. The pretty wood at our feet, with here and there a glimpse of the brawling stream, the sloping lawn around my cousin's house, and the wild rocks and woods, topped by purple moors, above it—all this lay right before us ; while to the left, bounded by hills greyer and yet more grey, stretched the widening valley of the Swale.

Emerging on the moor, the air was laden with the sweet perfume of the Ling. The rich purple was varied by patches of verdant green ; and, upon approaching one of these oases, I found two Ferns decidedly different from those I had yet become acquainted with, though the kidney-shaped spore-covers, testified to both belonging to the Lastrea family. One closely resembled the Spreading Shield Fern, but its leaflets were broader, and not curled in ; and its lower branches were less spreading. The leafy part of the frond was still triangular, but the base was narrower in proportion than in the former Fern, and the scales upon the rachis were blunter. Its characteristics answered to the description of the Spiny Shield-Fern (L. spinulosa, *Plate II., fig.* 3).

The other Fern was of the gracefully sloping contour of the common Prickly-Fern ; but its more upright form indicated a closer resemblance to the Male Fern, as did also its bluntly-notched pinnules. The spore-masses were arranged in a faultless line along the under margin of the

leaflets. Like the Male-Fern, it is bipinnate. A swee[t]
odour emanated from the plant, arising from numerou[s]
microscopic golden glands which covered the under sur[-]
face. All these particulars proved it to be the Heat[h]
Shield-Fern (L. oreopteris, *Plate II., fig.* 4). It forme[d]
the leading feature in a miniature landscape of exquisit[e]
beauty; for, as I knelt to gather some of the fragran[t]
fronds, I espied the coral Lichen clustering beneath it[s]
shade, like a gay parterre shadowed by delicate lime tree[s.]
The white fronds of the Lichen, with its red tips, an[d]
beyond, the horn-like branches of the Reindeer Mos[s]
contrasting with a soft purple cushion of wild thyme—
surely some such scene as this must have given rise t[o]
Mrs. Heman's description:—

> "Beneath these plumes
> Of waving Fern look where the Cup Moss holds
> In its pure crimson goblet, fresh and bright,
> The starry dews of morning."

It was a lovely picture, and I felt to grasp it more res[t]
fully, and to love it better, than the wide prospect [of]
wood and hill and valley beyond. Truly it is a weakne[ss]
of our mortal sense to imagine a thing trifling because [it]
is small. The vast peat mosses, which furnish fuel f[or]
thousands, are formed of an insignificant plant; whi[le]
mountain ranges are composed of skeletons of anima[ls]
imperceptible to the naked eye. To learn the value [of]
trifles we should study nature. In its lowest forms v[e]
see the vast importance of mere atoms; detecting [in]
microscopic Fungi causes of vast blessing, and of u[n-]
utterable dread. A tiny plant or insect testifies to t[he]

wisdom of its Creator, and raises the heart of the earnest-minded observer in adoring gratitude to Him.

A Cheshire friend sent us the Marsh-Shield Fern (L. thelypteris, *Plate II., fig.* 5), from some of the swamps in her neighbourhood. It is a smaller Fern than many of its family, and of a tender and succulent habit. Some of its fronds are barren, and some fruitful, and the latter grow taller and are more rigid in habit, exhibiting the proud consciousness of superiority, which we had already noticed in the fruitful fronds of the Oak Polypody and Parsley-Fern. The spore-masses are more distant from one another than in the other species. The cover falls off very early. The form of the frond is what we call linear-lanceolate—*i.e.*, narrow in comparison to its length. The pinnæ do not stand so thickly on the rachis as in the other Shield-Ferns, and there are no scales. It is difficult to keep it alive in a fernery, where all its brethren are easily naturalized; it needs a moist shady nook, but its succulent habit makes it a tempting prey to snails, which seldom neglect to attack its fronds when in search of a meal.

A neighbour contributed a Fern from Ingleborough— the Rigid Shield-Fern (L. rigida), another of the order with a very decided predilection for limestone districts. Its form is generally upright and sturdy. The stem is thick and scaly, and the general character resembles the spreading Shield-Fern, but it does not attain nearly so lofty a growth.

It was not till a later period that I made the acquaintance of the Recurved Shield-Fern (L. fœnisccii). It grew in a lane near Benenden, in Kent, and attained the height of something under a foot. Its general appearance might entitle it to be called the Parsley Shield-Fern, for each pinnule is curled outwards at the edges, giving the whole frond a crimpled appearance. Afterwards I found the same Fern in woods in Cornwall, and there it was more than a foot high. But the most beautiful specimens I ever beheld were in the Isle of Arran; there the fronds measured two feet and upwards. They were triangular, as in the Spreading species, the upper part of the rachis and the elongated pinnæ bending most gracefully. It was growing most luxuriantly at the entrance to some damp caves in the old coast line, to the left of Brodick Bar.

The Crested Shield-Fern we got from a garden. A sanguine fernist believed to have found this rare plant in the Bedgebury woods, in Kent; and indeed the narrow erect fronds, with the broader and more distant pinnæ, gave the Fern an exact resemblance to the form of the Crested species (L. cristata, *Plate II.*, *fig.* 6). But on subjecting the spore-covers to microscopic examination, it was found that their margins were notched, and this peculiarity attaches only to the Spreading and Spiny species, while those of the true cristata have plain margins. So the Fern was proved to be only a very marked variety of dilatata.

CHAPTER IV.

> " But open eyes may well discern
> Samples of pretty British Fern,
> Wall Rue, Spleenwort, Black Maiden-hair,
> On that old wall if scanned with care.
>
> " Then hasten, search the rocks and lanes,
> The meadows, brooks, the heather plains,
> The hedge, the dingle, copse, and all,
> But don't forget the old stone wall."

THE large and very interesting family of the Spleenworts comes next in botanical order. Here the spore-masses are placed in lines situated on the side veins, and the covers are flat and open towards the middle of the leaflet.

The alternate-leaved Spleenwort (Asplenium alternifolium, *Plate III.*, *fig.* 1), is one of the rarest in the family. It has been found in the south of Scotland, in Northumberland, Cumberland, and North Wales, but never in abundance. It grows more freely upon the black schist rocks between Conway and Beddgelert than anywhere else in Britain. It is a slender plant, the leaflets covering two-thirds of the rachis,

ASPLENIUM.

which is coloured with purple in the naked part. In a greenhouse it is evergreen.

The Forked Spleenwort (A. septentrionale, *Plate III., fig.* 2), has its fronds divided into three parts, and then forked. We could not find it among the Yorkshire rocks, and I afterwards sought it among the Braid and Blackford hills, near Edinburgh, but still in vain. It used to grow there, but the nursery gardeners have turned it to profit, and in so doing have exterminated it from its old haunts. I did get sight of some of its tufts, growing out of the basaltic columns called Samson's Ribs, in the Queen's Park; but it owed its safety to its impregnable position, which defied the attempts even of Fern dealers. Despairing of a wild specimen, I was obliged to have recourse to the nurseryman. The fronds are from two to three inches high, the rachis purple at the base, and the seed-masses much elongated. It is found on rocks and walls, in Wales, Westmoreland, and Yorkshire, as well as in Scotland.

The Rock Spleenwort (A. fontanum), has never gladdened our eyes in a wild state. I have seen it growing freely in fenceries, and a very pretty complete little cluster its fronds make. About three inches high, with broad, regularly indented leaflets, placed alternately nearly the whole length of the rachis, which is brown, and a dark brown root. The seed-masses here are rather oval than elongated, but they have the family characteristic of opening towards the middle of the leaflet. Its habitat is the highlands of Scotland and Derbyshire.

We started for a long excursion in search of more specimens, setting forward with hopeful hearts. The ride up that beautiful valley was enjoyable in the extreme. The hills on the opposite side were covered with purple Ling, across which cloud-shadows flitted with gliding motion. Keeping in a line with the river, we passed noisy brooks, whose waters were stained deep brown by the peat through which they had flowed, or were tinted with the paler hue of the limestone from the lead mines, where they had already performed the important duty of washing the ore. Whoever espied anything Fern-like was to call a halt, for the ponies were engaged to stop as often as we pleased, and for any given time.

We had passed through two villages, and reached a narrower part of the valley, when I espied Fern-tufts thrusting themselves from the crevices of a loose wall. We dismounted, and attacked the Fern with our knives; but though we quickly got fronds enough for our collection, it defied our efforts to procure a plant for the fernery. The ground on the other side was much higher than the road, and the Ferns were rooted in it, protruding only their long fronds through the openings between the stones. The rachis was a foot long, the pinnæ extending only about one-third of its length; the lower ones so broad and *pinnatifid* as to give a triangular form to the leafy part of the frond. The colour was of a full green, and glossy. The elongated spore-masses, nearly covered the backs of the leaflets, and the rachis was of a dark purple. The elegantly-tapering summit gave great grace to the Fern. It was the Black Maiden-hair Spleenwort (A. adiantum nigrum, *Plate III.*, *fig.* 3.) This was the first Fern that had ever attracted my attention. Many years ago, when visiting in Wiltshire, a friend of my hostess's came to see their guest, and brought a handful of this Fern, which she had gathered on the way. She gave her verdant bouquet to me, saying kindly, I don't know whether you are a botanist or not, but I feel eager to introduce the treasures of our neighbourhood to you." She understood Ferns and directed my attention to their beauties; and I then came to the resolution to study them whenever opportunity should offer. I have since gathered this Fern in Kent and Herefordshire, and very abundantly in Scotland, both in the Highlands and about Edinburgh.

In the same wall grew another Fern, much smaller, and familiar to my eye as the constant companion of the Ceterach. A rachis of from three to six inches in length, purple and very wiry; and oval indented leaflets, ranged on either side, bearing a row of elongated seed-masses on either side of the mid-vein, characterizes the Black-stalked Spleenwort (A. tricho-manes, *Plate III., fig.* 5). Every one who notices Ferns knows this one, the frequent inhabitant of rocks and walls. Here in Swaledale, it grew on walls and bridges, in old quarries, and from under gnarled roots in the rocky woods. Sometimes its tufts will spring from the lintel of an old barn; we found it in such a situation afterwards. Every county that I have visited has yielded me this pretty cheerful Fern; its very scent is dear in memory of woodland rambles.

As we proceeded, the valley became more wild. We passed through a very sequestered village, Gunnerside, and ascended some rising ground, from whence we had a splendid view of the wild hill-country stretching far away to the very borders of Westmoreland; and, by a little divergence, we caught sight of the pretty waterfall of Ivelet. The road led along the edge of one of the hills. We passed the mouth of a lead mine, and the miners whom we met greeted us with cordial goodwill, albeit, their manner, as well as that of all the country people in that district, partakes more of Saxon bluntness than of Norman courtesy. Honest, true-hearted people, we can dispense with surface culture for the sake of your staunch goodwill! A very rough road led down the hill. We crossed a romantic bridge which spanned the waters of the Swale: and, tying our ponies to a gate, we scrambled down a rocky wood, and arrived in due time at the foot of a deafening waterfall. The narrow gorge, shut in with rocks and wood, was wild and lovely in the extreme. The hills on either side were high, and the river seemed to have washed a passage for itself of upwards of a hundred feet deep: only a sturdy block of mountain limestone seemed to resist the further encroachment of the insidious waters, and so they were compelled to pour over it in the manner in which they were now doing. They indemnified themselves, however, for its interruption, by digging a deep hole immediately below it, into which they hoped some day to tumble it. Seated upon rocks there, we ate our sandwiches, drinking from a spring which bubbled from a bank to the left of us.

After our refreshment we began our search, and were soon rewarded by finding several plants of the Green-stalked Spleenwort (A. viride). This varies from the Black-stalked species chiefly in the colour of the said stalk; but the paler tint of the leaflets, and the less rigid habit of the frond, are also abiding marks of distinction. A few plants were found nestling in the crevices of dripping rocks: but the best locality for procuring it in that neighbourhood is the Buttertubs, great quantities of it growing about these curious chasms.

In returning, we took the other side of the valley, passing over broken ground plentifully adorned by the fragrant Butterfly Orchises, and coming out upon a road opposite to the mine that we had passed. Near Gunnerside we crossed the river again by means of a curious sloping bridge, unique in its style of architecture; and here we found a Fern resembling the Forked Spleenwort, but with the segments of an oval form, and toothed at the summit. Its dark green hue, short stature, and wiry stems are very like its brother above named. Gathering some of this, we continued our ride, ascending the hill and reaching the wide moor-pastures, now gay with waving Cotton Grass and the orange spikes of the Bog Asphodel.

We came to a quarry, or rather a series of quarries, on the steep hillside, and found that the stone was a conglomerate of the shells of the giant lima, called in the familiar language of the country, cockle. The stone was in a very decomposed state, and we found no difficulty in disinterring some of the heavy shells, the markings on which remained as perfect as when they were living. Smaller shells, closely allied, were there, and stems of animal lilies in abundance. The Cypress-like Ferns were not waving over these, as they waved over the corals in the wood, but the little Spleenwort, called Wall Rue, was resolved that their tomb should not be without verdure (A. ruta-muraria, *Plate III.*, *fig.* 6). This was the same Fern which we had gathered on the bridge; it resembles, both in form and colour, the leaf of its godmother, the garden Rue. It is very common in walls, being very abundant about Richmond, Edinburgh, Yatton, and indeed most places where limestone is used. But it, as well as its two brothers, the Forked and Alternate Spleenworts, is very difficult to cultivate. It will plant itself in a brick or other wall, and flourish despite a thick covering of dust and cobwebs; but you may take the greatest pains to coax it in the fernery, and yet fail entirely. You may plant it in fine sand, or in mould mixed with sand; you may shade it and water it; you may give it full sunshine, and let it be parched, or shade it with greatest care: but you cannot at all calculate on the probability of its life. I have known it carried to a pet fernery with the stones from the wall on which it grew, and be cemented among those stones again: I have known it flourish the first year—its year of difficulty, as one would suppose—and die next year, when every one would imagine that all was in its favour. If it have good drainage, and the bricks among which it was born, it may possibly live; without these it is sure to die.

It was vain to hope to find either the Sea or Spear-shaped Spleenwort in those inland dales; but at a subsequent period I had the pleasure of coming into possession of both. The Sea Spleenwort (A. marinum, *Plate III.*, *fig.* 4), I have found in caves at Dawlish, hanging from the roof like verdant drapery. The fronds varied in length from three to eight inches, of a vivid green; the leaflets of an irregular egg-shape, lobed on one side, and toothed along the margin, in substance thick and glossy. Afterwards I saw some plants of it among the rocks surrounding Looe Island, off the Cornish coast, and some fronds were given to me from the mainland. It is rarely, if ever, successfully cultivated in a fernery, but does very well in a Wardian case.

One little plant of the Spear-shaped Spleenwort (A. lanceolatum), was given to me in the south of Cornwall. The fronds were only three or four inches long; but I know that it often attains a much larger growth. The pinnæ extend for three-fourths of the length of the rachis, and taper to the base and summit, in the style that is called *lanceolate*. The leaflets are triangular in form and bright green, and the spore-masses are less elongated than in any other of the family, except the A. fontanum. To my great grief my treasured plant died, and I had no chance of procuring another, until, when staying in Kent, an Irish hawker came to the door with Ferns to sell. He had selected this mode of gaining a livelihood for the love of its irregularity; it was at any rate better than work. He was overjoyed at our appreciation of his Ferns, of the nature and character of which he was thoroughly informed, and he eagerly assured us that he would get us any Ferns that we wanted when he went his rounds, if we would give him the honour of a letter of commands. The address he gave for our expected communication was the most amusing part of the transaction—

"Patrick O'Leary,
Jolly Sailor Inn,
Mount Zion,
Tunbridge Wells."

All the members of this very attractive family haunt stony places, growing out of fissures in rocks and walls, from whence it is most difficult to take their roots uninjured. It is a pretty sight to behold these graceful plants lavishing their beauty upon the otherwise barren rock, or adorning the crumbling wall. It proves that God will leave no corner of His creation without its appropriate and harmonious beauties. The ancient rocks with their entombed organisms rejoice in new life, as the stone-shale and bright insects creep across them, and the verdant fronds of the Spleenworts kiss their aged surface at every motion of the air. The smiling verdure delights the eye, and brings to mind the words of sacred song—

"O all ye green things upon earth,
Bless ye the Lord,
Praise Him, and magnify Him for ever!"

CHAPTER V.

" Where the water is pouring for ever she sits,
And beside her the ousel and kingfisher flits ;
There, supreme in her beauty, beside the full urn,
In the shade of the rock, stands the tall Lady-Fern."

WE set ourselves one day to follow the course of the mountain stream from its junction with the Swale to its source in the moors. It was a difficult undertaking. Sometimes the rocks overhung the margin, and were so steep and high that it was impossible to climb over them; then our only means of progress was found by springing from boulder to boulder, and gaining the other side of the stream, and keeping along that bank until a patch of entangled brushwood interlaced with briars again stopped up our road, and we must recross, and pursue the side first chosen until again interrupted. Presently we came to where an accommodation road crossed the brook, by means of a bridge thrown from rock to rock. The archway formed the frame of a wild picture of rock, and waterfall, and drooping trees, with such a wealth of golden flitting lights, and deep shadows, as might have formed a rare prize for any artist. We passed under it, and from the sides

CYSTOPTERIS.

and top of the arch hung tufts of the Black-stalked Spleen-wort, and of a light feathery Fern, more like a miniature of the pinnatifid Lastreas. Eagerly gathering and examining some of these fronds, I found the spore-masses covered by a bladder-like envelope. These spore-clusters were round, and in many instances the white cover had disappeared. The fragile nature of the plants, the delicacy of their texture, and, above all, the peculiar spore-cover, pointed the Fern out as Cystopteris fragilis, the Brittle Bladder-Fern (*Plate III., fig.* 7). In the specimen in my hand the rachis was dark-coloured, and had a few scales towards the base. The frond was broadest in the middle, and tapering above and below, though only slightly decreasing in the latter direction. Each pinna contained several pinnules, which were toothed at the margin, and the spore-masses were borne on the branching veins. The plants growing outside the bridge were laden with spores; those under the shade of the arch were but scantily furnished with them.

Afterwards we found a similar Fern of more elongated form, and the pinnules were more sharply and decidedly cut. This was the plant by some considered a variety of C. fragilis, and by others accounted a separate species, C. angustata.

Another variety, rather toothed than cut—*i.e.*, with the points blunt and rounded, is also found among those Yorkshire rocks and old walls, C. dentata.

As we progressed higher and higher up the stream, climbing upwards towards the hill-country, we found one or other of these Bladder-Ferns among the rocks and boulders by the brookside. The fernery supplied another species, the C. Dickieana, or Dickie's Bladder-Fern. Its pinnules are broader, and the pinnæ placed more closely together. Its form is more compact than that of its brethren, it is a pretty Fern, and flourishes well under cultivation.

The Alpine Bladder-Fern resembles the Brittle species, but it is smaller and its pinnules more finely cut, and the middle vein is straight. As its name signifies, its habitat is alpine.

The Mountain Bladder-Fern (C. montana), resembles the Limestone Polypody in form, but is smaller and has more numerous leaflets, each of which is toothed, and the teeth fringed. It is a common Fern in high latitudes, and is only found in Britain at very high elevations.

After having passed the foot-bridge, and lost sight of our usual path, we came to a very wet bank, where a spring contributed its small amount of water to swell the stream, but without providing it with a channel, so that it diffused itself over the low ground by the brookside. I tried in vain to step only upon stones; I was obliged to trust to a cushion of moss, and my foot sank ankle-deep in water. But I could give no attention to the state of my boots, for immediately before me rose a tall group of Ferns, light and feathery in form, and bending most gracefully in every direction—their colour the most delicate green. The lance shaped contour, resembling that of the Male Fern, and the finely cut foliage, convinced me that it must be the Lady Fern. Shining drops of morning dew hung heavily on its tapering pinnæ and weighed them down. The plant was nearly two feet high, and the grace of its appearance was exquisite.

The spore-covers were fastened at one side, and the other edge was fringed; the masses were nearly circular, and the covering circular or kidney-shaped. The difference between this Fern and the Lastreas consists chiefly in their spore-covers being attached at the indentation, and those of the Lady-Fern (Athyrium Filix-fœmina, *Plate III., fig.* 8), at the side.

The plant and its position reminded me forcibly of Sir Walter Scott's description :

> "Where the copsewood is the greenest,
> Where the fountain glistens sheenest,
> Where the morning dew lies longest,
> There the Lady-Fern grows strongest."

As we proceeded along the wood, we found the Lady-Fern again and again. In a very boggy place she was growing in a narrower form, and the pinnæ farther from one another. I afterwards found that this was the variety named *Rhæticum*. It has sometimes a red stem, and a Fern-fancier of my acquaintance always calls it his "red-haired lady." There is a variety with the pinnules broader, which is called latifolium. It has only been found in the lake district. One of smaller stature and very delicate foliage, frequenting some parts of the western coast of Scotland, is called marinum. The Lady-Fern is as abundant in Ireland as the common Brake is here, and is used, like it, for packing fruit and fish.

Near an ascending path, which we were compelled to take, failing the possibility of progress close beside the stream, grew the well known Hart's-tongue. Clusters of the Prickly Shield-Ferns were growing by its side, and the contrast of form and tint between the two plants was perfect; the feathery pinnæ of the one, and the glossy

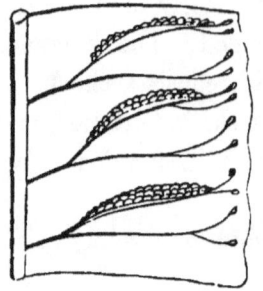

SCOLOPENDRIUM.

tongues of the other. Long narrow spore-masses following the course of the side-veins, and bearing coverings which split up the middle, are the characteristics of this Fern (Scolopendrium vulgare, *Plate III., fig.* 9).

The variety called *crispum* has the edges frilled, and is a very pretty plant for ferneries. The variety named Polyschides is also popular with Fern-fanciers; it is much narrower than the common form, and less graceful. We found some plants in the wood with the fronds forked at the ends.

The common Hart's tongue used to be valued as a medicine in England, and is still so in France and Scotland. The Male-Fern, too, and the Brake, were once used medicinally. In tropical climates the pith of Ferns is a general article of food, and there is scarcely one well-defined group, that does not boast an edible species.

We presently reached the top of the wood, and, climbing over a wall, found ourselves in a good footpath. Following this, and passing through gaps in the wall, called stiles, and certainly invented before crinoline, we entered a little copse, bordering on the grounds. Here, under the Birch trees, grew Ferns of an entirely different description to any I had yet seen. Three or four erect fronds, of about one foot to one and a half in height, rose from the centre of the plant; the pinnæ simple and narrow, and bearing a line of fruit on either side the midvein. All these pinnæ were turned inwards, so that in many of the fronds those ranged on either side the stem touched one another at the points. Numerous fronds with broader pinnæ, surrounding these, were bowed outwards, or lay quite flat on the ground. None of these fronds had any seed upon them; the old rule of the pre-eminence of the fruitful fronds was still in force! The long narrow seed-masses had coverings which opened towards the centre of the leaflets. This circumstance, with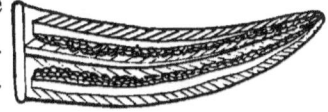

BLECHNUM.

that of the upright fruitful and recumbent barren fronds, proved the plant to be the Hard-Fern (Blechnum boreale, Plate III., fig. 10), so common in all our hilly districts, but welcome for its beauty wherever it occurs. It is sometimes called Fox-Fern, because of its odour, but it has no reason to feel flattered by the compliment.

Of course it was by no mere chance that I had found the Ferns so much in the order of their arrangements. I had shut my eyes to the species that I was not ready for. The Hard-Fern was in the wood at Kisdon, but I knew it was not yet time for it. The Lastreas grew side by side with the Polypodies, but I turned away from them as soon as I was convinced that they did not belong to the group I was in search of. I found this plan succeeded in keeping my mind clear.

But such a course as this would not act well in a district where there are but few Ferns. Then it would be best to take every variety you can find, and dry them, and when assembled bring your book to your specimens, use your lens, and patiently observe and compare. The plan of advancing step by step, rather than of charging a study full front, is one I would press upon all botanists. And it is the same in matters touching the heart ; if the path of life be thorny, don't look beyond the next step. We can see the depth of gloom before us, and yet catch no glimpse of the sunlight darting through the trees. To keep the mind clear, and the heart cheerful, we must hold fast by the counsel of the Omniscient Teacher—"Sufficient unto the day is the evil thereof."

1 Brake. 2 Wodsia. 3 Maiden. Hair. 4 Tunbridge Filmy Fern. 5 Bristle Fern. 6 Royal Fern
7 Moonwort. 8 Adder's Tongue. 9 Jersey Adder's Tongue

CHAPTER VI.

> " He orders the wind to sweep over the Brakes,
> Which rise and recoil like the billows of ocean;
> At His breath the leaf of the Maiden-hair shakes
> With the Aspen's tender and quivering motion.
> He dresses Osmunda in stately array,
> The Filmy-Fern covers with warm leafy shade,
> The Bristle-Fern frond He baptizes with spray,
> For o'er all creation His grace is display'd."

THERE are very few localities where we need seek the common Brake in vain (Pteris aquilina, *Plate IV., fig.* 1). In our Yorkshire rambles we found it in wood, and pasture, and moor. Sometimes its fronds were scarcely a foot high, at other times they rose to four or five feet. The roots grow very deep into the ground, which makes it very difficult to eradicate; but frequent mowing will kill it in progress of time. The rachis is only branched for about half its length : it is very strong and tough. The branches spread widely, and are set with pinnæ of a firm texture. The spore-masses are placed in a line upon the under margin of the leaflets, and covered by the rolled-in edge. When the fronds decay they form good manure, especially for potatoes. Sometimes this

PTERIS.

Fern is burned and its ashes used in the manufacture of glass, and in some places it is burned for fuel. There are districts where it is greatly in vogue as the litter of cattle. In a work on "The Channel Islands," by Professors Ansted and Latham, there is an account of another mode of using Ferns :—" Each cottage and old farmhouse has, in the kitchen or principal sitting-room, a wooden frame spread with dried Fern, called the 'lit de fouaille,' or Fern-bed, on which the inhabitants repose in the evening. This custom is, no doubt, French, and very old. It is connected with all the habits and traditions of the people, and comes into use on such occasions as the Vraic harvest, and on all festivals. The older people more especially resort to it ; and, though rough, it is by no means an unsightly piece of furniture. It corresponds with the chimney-corner in an old English farmhouse where wood is still burnt, and where pit-coal is an unheard-of-novelty." When the stalk is cut across, there is a marking not unlike a tiny picture of an Oak tree. Cut slanting the same pith-mark resembles a spread eagle. Its specific name, aquilina, is given on account of this fanciful resemblance. It was believed in old times that those

"Who gathered Fern-seed walked invisible,"

and people used to go out on St. John's-eve to collect it with great ceremony, as Leyden professes his intention of doing :

"But on St. John's mysterious night,
Sacred to many a wizard spell,
The time when first to human sight
Confest the mystic fern-seed fell ;

> Beside the sloe's black-knotted thorn,
> What hour the Baptist stern was born—
> That hour when heaven's breath is still—
> I'll seek the shaggy fern-clad hill,
> Where time has delved a dreary dell,
> Befitting best a hermit's cell,
> And watch, mid murmurs muttering stern,
> The seed departing from the fern,
> Ere wakeful demons can convey
> The wonder-working charm away,
> And tempt the blows from arm unseen,
> Should thoughts unholy intervene."

The Maiden-hair represents the last family of Ferns in the Aspidiaceæ group. Here the spore-masses are narrow, and curved into the form of a crescent. Our one native species, Adiantum capillus-Veneris (*Plate IV.*, fig. 3), is very rare, being only found wild in Cornwall, Devonshire, South Wales, Ireland, and the Isle of Arran off Galway. It is the most graceful of all the Ferns. Its stems are dark purple, slender enough to suggest the idea of hair, and quivering under the weight of the fan-shaped leaflets. My specimen came from Ilfracombe; but I had not the delight of finding it. The donkey-women make a monopoly of it, and sell it to all Fern-lovers. It was in vain to coax and wheedle, to promise a larger sum for the pleasure of gathering it myself. The woman who brought it at last dilated largely on the difficulty of reaching the spot

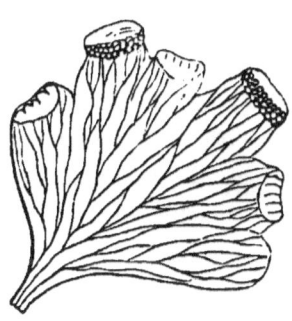

ADIANTUM.

where it grew : she had to climb precipices, creep through holes in the rocks, jump over chasms, and encounter dangers equal to those of "Arabian Nights" heroes. And when, making full allowance for her spirit of romance, it was argued that "what woman had done, woman might do," she assumed a different style of defence, and the other women would take her "poor dear life" if she divulged their secret.

A decoction of this plant is said to be good for promoting the growth of the hair. A friend of mine, sojourning for some months at Amalfi, near Naples, found her hair begin to fall off with the heat of the climate. At the same time she noticed the luxuriance of the hair of the Italian girls around her. She asked the maid who waited upon her what she used to make her hair so beautiful, and she said only the leaves of a plant infused in water. With characteristic politeness she hastened in search of the plant for her lady's use, and to my friend's great astonishment returned with a handful of the Maiden-hair Fern. A mixture called capillaire, in which this Fern is a principal ingredient, was formerly much used in England for this purpose. Certainly it is a most elegant plant for a house fernery, and is the prettiest possible addition to a bouquet, more especially if it be destined for a bride.

The order succeeding that of the Aspidiaceæ is called Hymenophyllaceæ. The Ferns belonging to this order have the spores in a cup-like receptacle, situated on the edge of the frond.

We sought in the Swaledale woods for the tiny Filmy-Ferns, but in vain. In a subsequent visit to Looe, in

Cornwall, I heard of a wood where the Tunbridge Filmy-Fern (Hymenophyllum tunbridgense, *Plate IV., fig.* 4). was to be found. We followed the course of the river first through hilly woods, and then along a shady road; then came a thicker and darker wood, with rocks here and there, and there was the wee Fern mingling with the Mosses, and scarcely distinguishable from them. The texture of the leaflets is nearly transparent, and the veins are very strong. The tiny cups containing the spores are seated on the leaf-

HYMENOPHYLLUM.

let near its juncture with the stem; their edges are unevenly jagged. I have since received beautiful specimens from the neighbourhood of Bridgewater.

Wilson's Filmy-Fern (H. Wilsoni) did not crown my search till later. I was rambling on the coast of Arran, and dividing my attention between the beach and its precious wreck, and the old coast line of rocks with their verdant inhabitants. Here it was that the Lastrea Fœnisecii was growing in such luxuriance; and in the very cave of which it was the gate-keeper I found Wilson's Filmy-Fern, clothing the moist sides as a curtain. Here the pinnæ turn to one side, and the tiny fruit-cups stand on minute stems, the margin being cut into two equal points. Passing through the village of Currie, with my arms full of these treasures, a "daft body" came up to me, and looking very kindly in my face, she asked, "Will these be Fairns, then?" I answered in the affirmative, and asked her if she liked them. "Na, na," she replied, "I've reckoned nought o' them; but I'd like a wee bit to mind me o' you, leddy." I gave her a frond of the large Fern, which she stroked caressingly, and then added a bit of the Filmy-Fern. "This is the least Fern of all," I said, "so you must have a bit of that too." "That a Fairn!" she exclaimed, "I shall ca' it the Fairy Fairn." Since then I have called it the "Fairy Fern" also.

A benevolent friend bestowed upon me a frond of the Bristle-Fern (Trichomanes brevisetum, *Plate IV., fig.* 5), It is the glory of the Killarney Waterfalls, growing freely in the style of the Filmy-Ferns, but only when within reach of the spray. The form of the frond reminds one of the Hare's-foot Fern of our greenhouses; the leaflets are deeply cut and lobed,

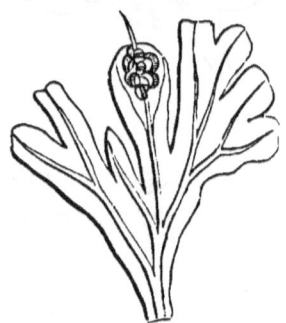

TRICHOMANES.

and their membrane extends along either side of the main stem. The spore-cups have a bristle growing from their base, extending beyond the margin: this gives the name to the plant.

The Royal Fern (Osmunda regalis, *Plate IV., fig.* 6), has an order to itself, Osmundaceæ. This noble Fern rises to a height of from two to five feet. The leaflets are of an elongated heart shape, and bright green colour. The fruit is in a cluster at the summit of the frond. There are generally only a few fertile fronds

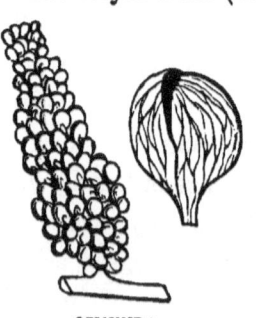

OSMUNDA.

to a greater number of barren ones. Marazion Marsh was the first locality where I found this beautiful plant in abundance. We were seeking plants of any and every sort in that first-rate botanical field. Plots were there covered with the delicate pink bells of the Bog Anagallis, and here and there spikes of the Musk Bartsia remained. Suddenly we beheld the compound spires and fresh green foliage of the Osmunda. In my eagerness I forgot the swampy nature of the ground, and plunged ankle-deep in water. I heeded not any such trifling inconvenience, for I had found the object of my ambition—the Royal Fern, or Osmund's Roy; and it was not only one plant, hundreds were there, growing under remnants of old wall and hedge. There, on the site of the old Jewish town, behind what may have been the carefully-planted fence of those ancient inhabitants of Marazion, those early miners and traders in Cornish tin, flourishes now the noble head of the English representatives of the still more ancient family—the family which flourished when rocks only a degree less ancient than those containing the ore were yet in course of formation.

Gerarde speaks of this Fern as "Osmund the Waterman," in allusion to an old tradition of a waterman living on the banks of Loch Tyne; and hiding his family among these tall Ferns during an incursion of the Danes. I saw the noble Fern again in a situation of more beauty, and scarcely less historical interest. It was growing freely on the banks of Loch Lomond, about a mile from Tarbet, reminding me forcibly of Wordsworth's description of it about English lakes:—

> " To point out, perchance, some flower or weed too fair
> Either to be divided from the place
> On which it grew, or to be left alone
> To its own beauty. Many such there were,
> Fair Ferns and flowers, and chiefly that tall Fern
> So stately, of the Queen Osmunda named ;
> Plant lovelier, in its own retired abode,
> On Grassmere's beach, than Naiad by the side
> Of Grecian brook, or Lady of the Mere,
> Sole sitting by the shores of old romance."

Our Yorkshire rambles supplied us with specimens of the last order of Ferns, the Ophioglossaceæ.

In the hilly field beyond our favourite wood we found the curious Adder's-tongue (Ophioglossum vulgatum, *Plate IV., fig.* 8). It consists of a broad sheathing leaf or frond, and a tongue formed of a double row of spores shooting up from its centre, somewhat in the style of the wild Arum. I have seen this strange Fern in Wiltshire some years before ; an old woman, a dealer in simples, had taken me with her to gather it. It was growing then, as now, in a pasture field ; but she knew its situation perfectly, and parted the long grass at exactly the spot where the sheath-like frond was standing. She concocted a kind of ointment from the plant.

In some other fields we found the Moonwort (Botrychium lunaria, *Plate IV., fig.* 7). It has a double row of crescent shaped dentated leaflets, each marked with a dark

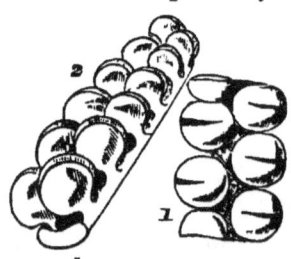

1. OPHIOGLOSSUM.
2. BOTRYCHIUM.

stain in the shape of a horse's shoe. This plant bears its seed in a branched cluster; like the Osmunda, it is used in medicine by village doctresses, and there is a superstition that it will open locks, and cause horses to cast their shoes.

The Jersey Adder's tongue (O. lusitanicum, *Plate IV., fig.* 9), differs from the common one in being smaller, and having its fronds narrower, more numerous, and not in a sheath.

Thus our collection is furnished with most of the members of the Polypodiaceæ order, characterized by naked seed-masses, and including the Polypodies proper, the Jersey Gymnogramma, Scaly Spleenwort, and Parsley-Fern.

The Aspidiaceæ order is also well represented, all its members having the seed-masses covered; Woodsia, with its fringed covers; Polystichum, with round covers; Lastrea, with kidney-shaped covers, attached at the indentation; Cystopteris, with its bladder-like covers; Asplenium, with its elongated covers, opening at the inner edge; Athyrium, with its kidney-shaped covers attached at the side; Scolopendrium, with its narrow covers opening in the middle; Blechnum, with its narrow covers opening along the inner side; Pteris, with its marginal line of seeds covered by the rolled-in leaflet; and Adiantum, with its crescent covers.

The three members of the Hymenophyllaceæ, or Urn-bearing Ferns are in our collection—Hymenophyllum, with naked cups; and Trichomanes, furnished with a bristle.

FERN ALLIES.

"Or with that plant, which in our dale
We call Stag's-horn or Fox's-tail,
  Their rusty hats they trim:
And thus, as happy as the day,
Those shepherds wear the time away."
                              WORDSWORTH.

CHAPTER VII.

FERN ALLIES.

FROM our childhood we have been familiar with various members of the Horse-tail family (Equisetum), their hollow stems being ever regarded as a prize in our juvenile games, because of their capability of being disjointed and joined again, and thus formed into chains or upright wands. Still they are not an attractive group, like their relations the ferns, and we need to look to their past history to learn the respect due to them.

In past ages the stems of Horse-tails rose, not like mere crops of reeds from the bog or river margin, but like huge forests of vegetable columns, grooved like the elaborate work of the stone cutter, and supporting the clouds on their lofty summits, thus turning the wide desolate waste into a vast cathedral. "The Brora coal," says Hugh Miller, "one of the most considerable oolitic seams in Europe, seems to have been formed almost exclusively of an equisetum, E. columnare."

The Horse-tail family are characterized by a distinct stem, furrowed, hollow, and with whorls of narrow pointed leaves, forming sheaths, protecting each joint in the stem, within which bristle-like branches are inserted : the seeds, or spores, are placed in cones, which grow on the summit of the stem, each spore being furnished with four spiral threads attached to its base, which curl and twist about in a very curious way, and move the seed along in various directions. The seeds shaken on white paper and damped, will be seen with the aid of a lens to be crawling about like as many minute spiders. The structure of the stem is very beautiful; the hollow centre is surrounded by a circle of pipes, and another circle of larger pipes encompasses that, the number of these larger pipes varies in different species, agreeing with the number of leaves in a whorl, these pipes run uninterruptedly through the stem, and you may suck water through them. A great deal of silex enters into the composition of the stem, giving it a remarkable roughness. The plants grow as weeds in watery places, and in arable land.

When setting forth in search of the various members of this family, the Corn Horse-tail (E. arvense) was the first to greet us. The fertile stems had risen from the ground, bearing large cones on their summits; their joints furnished with ample sheaths, but no branches. It grew among the young corn, in fields where the ground was heavy, ripening its seeds in April. The barren stems were in course of development, sculptured with ten grooves, and bearing from eight to ten branches in each whorl, the branches becoming shorter as the whorls neared the summit. According to Lightfoot, this plant "is troublesome in pastures, and disagreeable to cows, never touched by them unless compelled by hunger, and then bringing upon them an incurable diarrhœa." The root-stem of the Corn Horse-tail extends underground like that of the Brake Fern, making it very difficult to eradicate, and consequently as unwelcome to the farmer as the Colt's-foot or Butter-bur.

A certain "shore" in Kent, where the land is of a stiff clay, and the water stands all winter and spring, and great part of summer too, furnished us specimens of the rare Blunt-topped Horse-tail (E. umbrosum). Very early in the spring we took our walk along the Junction Road, the great high road between London and Hastings, and as we wandered along, seldom encountering any vehicle, so entirely had the railway absorbed the traffic, our attention was attracted by thick stems springing from the wet "shore," the deep sheaths covering the whole length, and overlapping one another towards the apex, which bore a large brown cone. Later in the season, when the Furze and Broom were flaunting their crowds of golden blossoms, and the Loosestrife was bordering every sluggish stream with its fringe of crimson spikes, then that "shore" became a perfect forest of Equisetum, each barren stem resembling a Fox's brush, so crowded were its whorls of branches. The shortening of the branches towards the summit, their upward growth, and vast number, rendered the top of the "brush" flat—the cones had disappeared months before.

The Great Horse-tail (E. telmateia, *fig.* 2) is the best living representative of the chiefs of the family in the Brora age. On the margins of ponds and sluggish streams, this plant grows to a considerable height, throwing up its cones first, and then replacing the naked stems with a forest of closely-packed clustering branches, among which the frogs croak harmoniously on damp evenings, and deposit their transparent eggs, while later, myriads of tadpoles wriggle round the half-submerged Equisetum stems. In ponds about Hawkhurst, in Kent, in ditches in Durham, and on the margin of the lake at Longleat, Wilts, we have gathered fronds of this Horse-tail between two and three feet in length. Haller informs us that in Rome the lower classes used this species as food, but cattle always avoid it.

As the Great Horse-tail claims superiority in size over its fellows, so does the Wood Horse-tail (E. sylvaticum, *fig*, 1) claim the superiority of beauty. Springing in shady well-watered nooks, in subalpine situations, this plant forms a graceful object in a beautiful landscape. In Swaledale we found this plant first; it abounds in thickets along the hill-sides between Melbecks and Summer Lodge. Growing about a foot and a-half in height, its whorls of branches becoming shorter towards the base and summit, each branch bearing clusters of branchlets, and bending downwards with a graceful curve, it resembles some elegant oriental tree, a palm or a cycad, much more than one of our despised Horse-tails. In our May ramble we found the plants in full beauty, each graceful stem surmounted by a small cone.

The Smooth Horse-tail (E. limosum, *fig*. 3) frequents similar situations as the Great Horse tail; we have gathered it in Swaledale. Sowerby tells us that it is "an active agent in the conversion of pools into swamps and morasses, which it abandons as the soil becomes elevated, so as not to admit the retention of water on the surface." We saw it performing this part in a shallow pond at Hawkhurst, which is now dried up altogether.

In this species the stem is smooth, the cone ripens in July, and then falls off.

Swaledale furnished our specimens of the Marsh Horse-tail (E. palustre), its stems are deeply furrowed, and it has from four to eight branches in a whorl, the branches becoming much shorter as they near the summit. It is a slenderer plant than any of the others which we have described, its whorls more distant, its branches fewer and slighter in form, the cone dark coloured and small, and the tint of the whole plant a darker green. There is a variety bearing cones on the upper branches, called Polystachion, and one with naked stems, called Nudum, but we have not found either of these.

On the borders of mountain streams in the Yorkshire dales we have occasionally met with the naked stems of the Rough Horse-tail or Shave-grass. The stems are of a glaucous green, very rough from the presence of a large quantity of the silex crystals, minutely striated, and bearing a small cone. This plant is imported from Holland for polishing wood, ivory, and metal; it forms a natural file. Mr. Baird, in his "Flora of Berwickshire," tells us that "the dairy-women of Muiroick and Chipchase, where the plant is plentiful, use it for smoothing their milk-vessels."

The Quill-wort (Isoetes lacustris, *fig.* 7) is a curious plant, and also ranked by most authors as a Fern Ally. Here the seed is contained in vessels embedded at the base of the leaves. It always grows submerged. Our specimen came from the Westmoreland lakes.

The Pill-wort is an equally curious plant, bearing its seeds in round hard balls attached to the stem. The habit of the plant is creeping, it bears clusters of awl-shaped leaves, and covers the surface of mud under shallow water. We have not succeeded in finding it.

The second great family of Fern Allies, the Club-mosses (Lycopodium) are much more generally interesting than the Horse-tails. Here the seed-cases are borne amid leaf-like scales, either mingled with the leaves or forming a cone. We were making a charming excursion in Swaledale when we first found the Common Club-moss (L. clavatum, *fig.* 4.) We had crossed the Butter-tub Pass, and were proceeding along Stag-fell, when the trailing stems and uplifted cones of that king of mosses first caught our eye. We gathered long branches, and did not wonder at the fancy of Wordsworth's shepherds for coiling it round their hats. Mr. Baird tells us that "the seeds of this plant are used in Germany for producing artificial lightning on the stage, for when dispersed in the air they may be ignited in the same manner as powdered rosin." He also informs us that "woollen cloth boiled with this plant acquires the property of becoming blue when passed through a bath of Brazil wood."

This species is common in all our mountainous districts.

The Interrupted Club-moss (L. annotinum) grows in Wales, and on the highest of the Scotch mountains, but we have found no specimens except in Switzerland. In this species the branches grow upright, while the main stem creeps, and the cones are thicker and shorter, and placed singly in the stems, not in twos and threes as in the common species.

For abundant specimens of the Savin-leaved Club-moss

(L. alpinum, *fig.* 6) we have to thank the Swaledale moors. It was the twelfth of August, a Sicilian vespers for the grouse, and we had arranged to meet the sportsmen at noon, and carry them the refreshment which they so sorely needed. Of course it rained—when does it not rain among the hills on the day when out-door entertainments are arranged? Of course it rained, and of course we defied the rain. Such a collection of cloaks and hats, and such contrivances for keeping the paniers dry, which the patient hill pony was to carry, along with the worst walker of our party! The strong British resolution of each was bent on keeping our tryst; but our energies were sufficiently tasked in making our way, up-hill and down-dale, amid heath and moss, or over rough stones, to preclude the possibility of conversation. On we plodded in single file, no cloaked and hooded figure attempting to entertain or be entertained. Then it was that the quadrangular stems of the Savin-leaved Clubmoss arrested my attention, but I gathered its tufts of branches, admired its plentiful cones, and proceeded on my way, too wet and weary to proclaim my good luck. But when we reached the rough shed on the moor, and a peat fire had dried us without quite choking us with smoke, the matter assumed a more cheerful aspect. Then I produced my Club-moss, sharing it among the plant lovers, and its four rows of blunt leaves, its pale glaucous tint, and the abundance of its fruit, received a full measure of interest and approval. Mr. Sowerby quotes an interesting passage from Sir W. J. Hooker's "Tour in Iceland," relative to this plant. "A vast heap of Lycopodium Alpinum lying before the priests' house drew my attention, and on inquiry I found it was used for the purpose of giving their wadmal a yellow dye, which is done by merely boiling the cloth in water with a quantity of the Lycopodium, and some leaves of the Vaccinicum uliginosum. The colour imparted by this process, to judge from some cloth shown me, was a pale and pleasant, though not brilliant colour."

The Fir Club-moss is a very marked species, growing in sturdy little shrubs on bare ground, like young fir-trees. This also flourishes upon the Yorkshire moors. There is a gloomy tarn upon the top of Summer Lodge Bank, its bed is formed of the *debris* from an exhausted lead mine, and its shores for some distance inland are heaped with the dark limestone refuse; there is a lovely prospect of grey hills in the distance but the near view is gloomy enough. But here, among the grey shingle, beside those cold still waters, numerous dwarf bushes live and flourish, brave little Fir Lycopods. In this species the seed vessels are situated among the leaves, over the whole length of the branches (L. selago) Pliny mentions that the Druids gathered the Selago with much ceremony, and used it as a cure for complaints of the eye, and as a charm to avert misfortune. Mr. Baird tells us that some of the Highlanders use the Fir Club-moss instead of alum in dyeing.

The Marsh Club-moss (L. inundatum) is one of the smallest of the family. We have only once been in its native home, or rather in one of its native homes, for it is not a very uncommon plant. It used to flourish on Rudd Heath, Cheshire, and still flourishes there in all probability, if the rage for drainage has left any part of that splendid botanical ground in its natural state. The leaves are long and pointed, with two sharp shoulders and a broad base; the cone is very large in proportion to the rest of the plant.

Our specimens of the Lesser Alpine Club-moss (L. Selaginoides) were sent to us from Blair-Athole ; we have none of us found it ourselves. It is more slender than the March species, but much more branched, and its general aspect is more like that of a real moss than one of the sturdy Lycopods. But if it seems slight and frail compared with the long tough stems of the Common species, and the sturdy bushes of the mimic fir-tree, how much more startling is the contrast with its giant ancestors of the coal measures, the Lepidodendra, which Hugh Miller describes as " great plants of the Club-moss type, that rose from fifty to seventy feet in height." A valuable homœopathic medicine is prepared from the Lycopodium.

The Spanish moss which forms beds of such intense and exquisite greenness in the conservatories at Kew, is a Lycopod, and other species there to be seen, delicately tinted with rose colour, or blue, belong to this family.

This group of fern allies completes the first section of the great tribe of Flowerless Plants ; these, together with the mosses and their allies, the Scale mosses or Liverworts, belong to the first or foliaceous division of Cryptogams ; while the Seaweeds, Lichens, and Fungi form the Aphyllous or leafless division.

RAMBLES
IN
Search of Flowerless Plants.

CHAPTER VIII.

MOSSES.

" The night is mother to the day,
   The winter to the spring.
And ever upon old decay
   The greenest mosses cling.
Behind the cloud the starlight lurks,
   Through showers the sunbeams fall :
For God, who loveth all his works,
   Hath left his hope with all."

HE next order of Flowerless plants to the Ferns, is the Mosses, a very large group, freely diffused in all the countries of the World.

There is no distinct flower in the moss, though the organs of fructification are of two kinds; that in which the seeds are formed is the most conspicuous, and is called an *urn*. This urn is covered with a veil during its immature period; the veil falls off before the seed is ripe, and the urn remains closed by a lid. When this lid comes off, the seeds are ripe, and are found arranged round a central column within the urn. The rim of the urn is bordered by sets of teeth; one set appears to belong to the outside, and one to the inside. The urn generally grows from a fleshy tubercle (apophysis), the station of which is generally at the base of the flower-stem.

The secondary kind of fructification is only present in some mosses; it is formed of membranous cylindrical bodies clustered in the axils of the leaves; they open irregularly at the point, and discharge a sticky fluid.

Mosses are among the first plants that spring up on the surface of inorganic matter; at first they appear like a green stain, merely consisting of granulating seeds, but soon clothing themselves with leaves, and then by their decay producing the first deposit of vegetable matter with which the soil is fertilized.

The large group of Mosses, including several hundred species, are divided into two great sections : first, the *summit fruited*, where the fruit stem rises from the end of the branch; secondly, the *side fruited*, where it rises from the side of the branch: but to these rules there are so many exceptions, that we need rather to direct our attention to the particular distinctions of the several groups.

The first group of summit fruited Mosses is the Andræa group. It is characterized by one great peculiarity, which has led scientific men to place it in an order by itself. Its urn splits into four valves. These Mosses are inhabitants of Alpine or sub-Alpine districts. They are small plants, with blackish red foliage, which gives them a burnt up appearance. The leaves fold over one another in eight rows.

Specimens of the Alpine Andræa have been given to us from Ben Nevis; the leaves are oval and pointed.

The Rock Andræa we have from Blair Athole. It is a commoner species, growing in short loose tufts, the stems a little branched, and its leaves blunt (A. rupestris, *Plate V., fig.* 13.)

The Black Andræa (A. rothii) is still darker in hue, and a little taller; the leaves are awl-shaped.

The Tall Andræa (A. nivalis) is of slender growth, and is only found on the margin of perpetual snow.

We went to the moors above Summer Lodge Bank in search of the second group of the order, Bog Mosses (Sphagnaceæ.) We reached the neighbourhood of the gloomy tarn, and began our search among verdant patches of swampy ground, while all around and amongst the ling the soil was covered with a thick carpet of white moss, varied with a pink hue. We soon collected several specimens of Bog Mosses (Sphagnum.)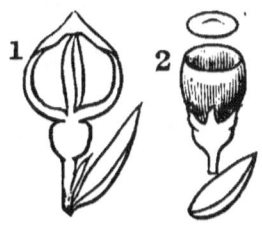

1. ANDRÆA. 2. SPHAGNUM.

These Mosses are characterized by having their branches arranged in clusters, their roundish urns on very short footstalks, and by having no proper roots. As this group seems to be designed by Providence to fill up water pools in bogs, roots are unnecessary. The densely crowded stems have little threads attached to them, by which they imbibe moisture : as the plant grows, the under part decays and deposits the used-up portions, while the acids, set free by the decomposition of its parts, uniting with that given out by other bog plants similarly decaying, forms a kind of tannin, which preserves the substances imbued with it, and renders them impervious to decomposition. This accounts for the perfect state in which bog oak, horns and bones of animals, &c., have been found in the Irish peat. The Bog-moss continues growing till it rises far out of the water, other plants twine amongst it, earth is gradually formed, and the barren swamp becomes in time a verdant place where cattle may feed. Who can tell but that, in centuries to come, yonder black tarn may be carpeted and covered by this useful and redeeming moss?

The red-tinted branch was the Blunt-leafed Bog-moss (S. cymbifolium, *fig.* 1 in cut); its little clustering branches are short, and its leaves concave and blunt. It is one of the largest of the family. Another, with pointed leaves, proved to be the equally common Slender Bog-moss (S. acutefolium, *fig.* 2). It is as tall as the former, but of a frailer form. It is also white, and tinged with pink. The Red Dwarf Bog-moss (S. rubellum) I found in great beauty on the hills above Oban. The ground was very marshy, so much so as to be dangerous for explorers. The stems of the moss were short, branched, and closely matted; the colour very red, and the leaves blunt. Another Bog-moss, with paler foliage of a straw colour, and with less crowded branches, was intermixed with the Red Dwarf species, and proved to be the Pale Dwarf Box-moss (S. molluscum). The leaves in this species are a broad oval. The Compact Bog-moss (S. compactum), with its forked stems, short crowded branches, and oval leaves, was sent to us by the most patient and successful of Moss collectors, Miss M'Leeray of Jude, Blair Athole. She also procured the fringe-leaved species for us (S. funbriatum), resembling the Slender Bog-moss in general appearance, but with the points of the branch leaves turned back.

From the margin of a peat pool we gathered a quantity of soft green moss, its long branches mingling with those of the Cranberry plant. This was the Wavy leaved Bog-moss (S cuspidatum, *Fig.* 3); when growing on dry ground it is generally white, but has a bluish green tinge at the tips, instead of a pink or lilac one; growing in water it is very attenuated, and of a brilliant green. Its clusters of little branches are always more slender and inclined to droop than those of the others we have here. The Spreading Bog-moss (S. squarrosum) is a very large species, firm, robust, and branched; its leaves spreading, turned back, and pointed. We have not found a specimen.

The group which immediately succeeds that of the Bog-moss is the Earth-moss group (Pascæ). What the Bog-mosses are to marshy places, such are the earth-mosses to clay banks, covering them with a green crust at first, which develops into tiny plants; these attain maturity very rapidly, sow thier seed, and die away in a very short time, thus scattering over the tenaceous soil a deposit of organic matter, which prepares it to nourish plants of a less minute organization. In the Earth-moss family the capsules have little or no stalks, the leaves are generally in eight rows, and the whole plant is wonderfully small. The urn is roundish, with a pointed veil splitting up one side, and no proper lid. We were too late in the year to find the Pointed, the Tall, or the Awl-leaved Earth-moss, and the other species do not attain maturity till autumn or winter.

The Straight-necked Earth-moss (Phascum rectum) we have from near Southport; it is exceedingly minute, with chestnut-coloured urns, and tiny crowded leaves. By far the most frequent member of this family is the Awl-leaved Earth-moss (P. sublatum, *Plate V., fig.* 1). It has a tiny stem, its leaves are lance-shaped and concave, and its urns pale brown. Great numbers of plants grow crowded together in dense patches, their foliage of a yellowish hue and the sharp points of their abundant leaves giving a hairy aspect to the plot. It grows on

sandy and clayey banks, and perfects its fruit in April.
Edward has it from Kent, and I from Wiltshire and
Herefordshire. Twenty plants
grow together in a patch the size
of a sixpence.

The Alternate-leaved Earth-
moss is common in the spring
upon fallow clay-ground : it has
a branch rising higher than its
capsule, or urn; the Serrated
Earth-moss, the Clustered and
Strap-leaved Earth-mosses are exceedingly minute and
evanescent.

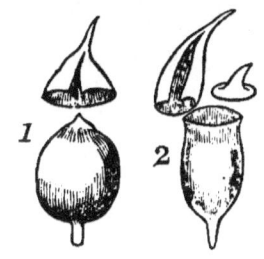

1. PHASCUM.
2. GYMNOSTUMUM.

The Weissias form the next group ; it includes the
Beardless Mosses, the true Weissias, and the Streak
Mosses.

The Beardless Mosses (Gymnostomum) have no fringe
on the margin of the urn, hence their name. They are
small plants growing on rocks, or earth, in thick tufts;
both the lid and the veil are beaked.

The Curve-beaked Beardless Moss (G. curvirostre,
Plate V., fig. 2).

Then there are the Small-mouthed Beardless Moss, and
the Spreading Beardless Moss, and the Curly-leaved
Beardless Moss, all of which are characterized by the urn
being contracted at the opening.

The Rev. Hugh Macmillan, in his "Footnotes from the
Page of Nature," tells us that Solomon's "Hyssop on the
wall" is identical with the little Beardless Moss, Gymnostomum truncatulum. He adduces the authority of Has-
selquist, who called it Hyssopus Solomonis, and described

it as growing on scattered tufts on the walls of Jerusalem.

The Weissia family have an oval erect urn, a lid with a sloping beak, and a simple fringe of sixteen teeth. They are small, with leaves in eight rows, and the stems grow in clusters.

Upon hedge banks all about in Swaledale we found large patches of fine, close, fresh green moss. A plant which we examined with a pocket lens, showed its leaves turned in at the edge, the lower ones lance-shaped, the upper narrower; the little oval urn was brown and shrivelled, for its fruit had been ripe two months before. This was the green tufted Weissia (W. controversa, *Plate VI., fig.* 1). A species with taller stems and crisped leaves proved to be the Bent-leaved Weissia (W. cirrhata, *Plate V., fig.* 2); and one with the edges plain was the Curly-leaved (W. crispula).

The Whorled Weissia (W. verticillata) grows on dripping rocks; we found it afterwards half encrusted with lime. It is a larger species, growing in a thick cluster, and with foliage of a vivid green.

Some plants of the Bristle-leaved Weissia (Brachyodus trichodes) were sent us from Castle Howard, they are as small as those of the Earth mosses, and even more delicate in structure, Don's Bristle mosses, (Anodus donianus, *Plate VI., fig.* 3) is the first member of the Bristle moss group, the characteristics of which are a wide-mouthed roundish urn, a large beaked lid, a small veil, single fringe, and bristle-shaped leaves. Don's Bristle moss vies with the Bristle-leaved Weissia in minuteness; it differs from all the other Bristle Mosses in having no fringe at the mouth of the urn. My specimens are from Longleat Park, Wilts, where they are growing upon a sandstone rock.

An important group succeeds the Bristle mosses, namely, that of the Fork mosses: it includes seven families. The first family was discovered by the botanist Blind, of Munster, and is hence called *Blindia*. (*Plate V., fig. 3.*)

The Northern moss family has but one British representative (*Arctœa fulvella*). The Dog's-tooth moss resembles the Bent-leaved Weissia, but its urn is shorter, and its leaves are keeled at the base: it grows in hilly countries. The next three families, though bearing different botanical names, are all Fork mosses in honest English, so we will discuss them as one great family.

The true Fork-mosses (Dicrannui) are very numerous. The many fruited Fork-moss (D. polycarpa) we have from the Highlands, where it grows on rocks. It has long stems, spreading lance-shaped leaves, and urns not much longer than broad.

The transparent Fork-moss (D. pellucidium) grows in a loose tuft, and is of a light green colour, and more slender urns, the fruitstalk a little bent at the neck. It grows in wet places. Our specimen was from dripping rocks near Richmond.

The Drooping leaved species (D. squarrosum) was contributed by our friend from Blair Athole, its leaves spread in every direction, and are turned back, the stems are forked, and its loose tufts are of a bright green. When it bears fruit the stems are only one or two inches high, but it grows much taller when barren.

We found the Lily Fork-Moss (D. heteromallum, *Plate V., fig.* 5), abundantly on a moist bank in one of the birch woods, and we have found it in every county that we have since visited. It grows in large patches, covering the earth with its yellow green silky foliage, and chestnut urns. When examined with a lens the leaves are seen to be toothed towards the narrow point; the fruitstalk is bent where it joins the urn, and the lid is prolonged into a beak.

The sickle leaved Fork-moss (D. falcatum), we have from the highlands, the leaves are all turned to one side, and bent, so as to give the form which characterises the plant. Its colour is dark green, its stems branched, and the lid beaked.

The Swaledale woods afforded abundance of the Broom Fork-moss (D. scoparium, *Plate V., fig.* 4), growing in extensive patches. A large handsome moss, with yellow-green, glossy foliage, broad at the base, but narrowing to a long fine point, with saw-like edges; the urns nearly erect, chestnut coloured, and with a beaked lid, the veils being beaked also.

The Marsh Fork-moss (D. palustre) closely resembles this, but the leaves are less narrow at the point, and the fruit grows more abundantly, several urns rising from one plant.

The Tall Fork-moss (D. majus), was sent to us from Blair Athole, it resembles the two last, but is still larger; it has clustering fruit-stalks like the Marsh species, and the foliage turned to one side, as is often seen in the Broom Fork-moss.

There are many other species in this family, but they have not rewarded our search.

The White-leaved Fork-Moss (Leucobryum glaucum) we did not find till a later period. When staying in London, I joined a pic-nic to Virginia waters. Under the trees there, within sight of the lake, I came upon a carpet of soft whitish moss. Its colour reminded me of the Sphagnum upon our dear Yorkshire moors. The stems were short—two or at most three inches in height, —slightly branched. They were stiff in form, and closely covered with lance-shaped, channelled, blunt leaves. The urns were short, and the footstalks short also; the latter were withering, their time of perfection being March.

The Purple Fork-moss (Ceratodon purpureum, *Plate V.*, *fig.* 7) is very common on waste ground, banks, fallow fields, and such places. Edward first found it on heaps of sand at Hawkhurst, being attracted by its abundance of shining purple urns, and fruit-stalks of the same colour. The foliage is of a dull green, the leaves spreading, much twisted when dry. I have since found it both in Yorkshire and Herefordshire.

The Swan-neck Mosses are near allies of the Fork-mosses. The Beaked species (Dicranodontium langirostre) has very long, narrow-pointed, curled leaves of a bright dark green. The fruitstalk bends when moist, concealing the urn among the foliage, but becomes erect when dry. It is a rare moss, very rare in fruit. Our specimens came from near Lennox.

The Compact Swan-neck Moss (Campilopus brevipilus) is a very attactive species. Its foliage is silky, the brown hairs on the lower part of the stem giving a dark expression, while at the tips of the branches the tint is of a golden olive. The leaves clasp the stem, and taper to a fine point. Kincardineshire is the only habitat of this moss that I know.

The **Bristly** species and the **Rusty** species (C. longipilus and flexuosus) inhabit wet rocks in subalpine districts. The **Dwarf** one (C. tarfaceus) is slender and clustered, and is found on ditch-banks, &c., with small Fork-mosses.

CHAPTER IX.

"The tiny moss, whose silken verdure clothes
The time-worn rock, and whose bright capsules rise
Like fairy urns on stalks of golden sheen.
Demand our admiration and our praise,
As much as cedar kissing the blue sky,
Or Krubal's giant flower. God made them all,
And what He deigns to make should ne'er be deemed
Unworthy of our study and our love."

 CHARMING quality in this Moss order is the power of revivification in the plants: pieces that have been dried and laid away for years still retain their vitality, thus rivalling the snail in the British Museum, which, having been cured and glued to a slab for years, found one happy morning that the glue had given way, upon which it stretched forth its horns, as if after a long, long sleep, protruded its broad foot, and had travelled half over the case when its movements attracted the eye of the curator of the department.

These old dried mosses, the precious gleanings from an old herbarium, when floated in water, expand to their original size and form; the minute cells of which they are formed fill again with fluid, and only their paler hue shows that they were not gathered yesterday. We floated them out on a rainy day, and selected one which belonged to the order succeeding the Fork-mosses for our first examination. The leaves were spreading, oval and pointed, the urn wide-mouthed, and the lid swollen and slightly beaked. The whole plant measured less than a quarter of an inch. Several plants were clustered together; the leaves in five rows, the upper ones crowded, the lower more distant; the roundish urn and convex lid made me believe it a Pottia, so called from the German professor, Pott, the first person who studied this family. My cousin came in and settled my doubts by assuring me that it was the Common Pottia (Pottia truncata, *Plate VI., fig.* 4).

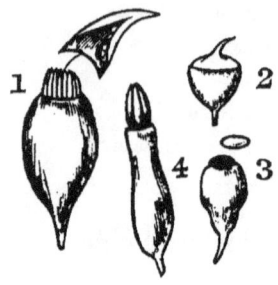

1 WEISSIA. 2 POTTIA.
3 PHYSCOMITERIUM.
4 PETRAPHIS.

The book informed us that there are an Oval-leaved Pottia and a Dwarf Pottia, an Oval-fruited Pottia, a Bristly Pottia, and a Lance-leaved Pottia, all of which frequent mud, sandy banks, or fallow ground. All these Pottias are without fringe at the mouth of the urn.

A specimen of Starke's Pottia (Anacalypta Starkeana), one of the species gifted with a fringe, was my reward for services on another occasion; it much resembles the Common Pottia: its redder and more oval urn, with the blunt lid and the dark or yellowish green of the leaves, distinguish it.

The group Trichostomæ, or Hair-mouthed mosses, contains four families: the Two-ranked mosses, the Twin-toothed mosses, the Hair-mouthed mosses, and the Screw mosses.

The first family have single fringes, containing sixteen

teeth, and the central column extends beyond the mouth of the urn.

The broad-leaved species is an Alpine moss, while the Thick-ribbed prefers the sea-coast.

The Five-leaved and Oblique-fruited Two-ranked mosses have their leaves distinctly in two rows: the former is a delicate moss, with a reddish urn. My cousin had put a specimen of it into the book for me (Distichium capillaceum, *Plate V., fig.* 8).

The Twin-toothed mosses have many of the teeth divided, hence the name. Their stems grow in clusters, their leaves are more or less lance-shaped, and have dots on the surface. There are a reddish species, growing on walls: and a dusky species, favouring limestone: and a Slender fruited species, frequenting the neighbourhood of waterfalls; and a Bent-leaved species, growing on elevated moors (Didymodon, *Plate V., fig.* 10).

The true Hair-mouthed mosses have their leaves in five or eight rows and the nerve reaches to the point of the leaf; the oval fruit is generally straight and dull, and placed on a long fruit-stalk.

Edward found the (Trichastomum Homomallum, *Plate V., fig.* 9,) Curve-leaved Hair-mouthed Moss upon a sandy bank in Kent. The earth had been freshly thrown up into a heap, and this Moss was the denizen of the new soil. It has short clustered stems, and leaves awl-shaped at the base, and dwindling to a silky point. There are sixteen pairs of teeth in the fringe, and they are sometimes joined together.

On a sandy rock in the same district he found the Twisting species, (T. tortile), with its half prostrate stem and spreading leaves, broader than in the last-mentioned species.

The Rigid-leaved Hair-mouthed Moss has erect leaves and longer fringe (T. rigidulum) ; and the Curly-leaved species grows near the sea. The glaucous one is peculiar to mountains.

Rambling along the road up Swaledale, we found various Mosses growing upon the curious bridge opposite to Gunnerside. Some of these had long, narrow urns growing on erect fruit-stalks, and upon examining them with the lens, we found that the teeth of the fringe were long and slender, and twisted round the pillar in the centre of the urn ; the lid was long and beaked. These being the characteristics of the Screw-moss family (Tortula), we set ourselves to determine the species. One had oval leaves, with long hair like points, the plants grew thickly together, forming a little cushion, and the abundant hairy points gave it a downy appearance ; the leaves seemed to have a thickened border, but this was only with the edges being turned back. This was the Wall Screw Moss (T. muralis, *Plate VI., fig.* 5), common on rocks and walls, throughout the kingdom.

Another species with pointed leaves, spreading in a starry shape, and of a bright green colour, was rendered remarkable by its very long slightly curved urns. Altogether it was a brighter, cleaner looking plant than his brother, it was the Awl-leaved Screw Moss (T. subulata, *Plate VI., fig.* 6).

Upon a bank near a quantity of the Fallacious Screw Moss was growing, seeming to enjoy the limestone rubbish. It was taller than the other species, and the leaves were of a dull green, twisted and turned back. There was no fruit on our specimens. The urns are reddish and oval (T. Fallax, Plate VI., fig. 7).

Early last spring, when wandering in the beautiful combes of Somersetshire, I found large cushions of the twisted Screw Moss (Tortula tortuosa, fig. 8). The stems were high and closely matted together, and the long leaves growing in dense whorls round the stem were all curiously twisted. The pale tint and elastic feeling of the cushions attracted my attention. Afterwards I found the same Moss upon rocks in Swaledale, but in both instances there was no fructification. I have recently received specimens from Blair-Athole with abundant urns, both them and the stalks yellowish-brown.

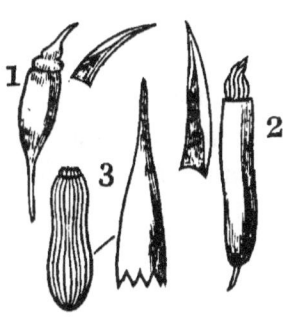

1. DICRANUM. 2. TORTULA. 3. ENCALYPTA.

Another Screw Moss I have found abundantly on trees and thatch. The great Hairy Screw Moss (T. ruralis, Plate VI., fig. 9), has tall stems, leaves spreading, turned back with hairy points, and long curved urns. In Yorkshire and Herefordshire it frequents trees, as also in the Bath neighbourhood; but the finest specimens I ever gathered were off thatched cottages in Wilts and Somerset. It is said to be a great preservative to the thatch.

Müller's Screw Moss (T. Mülleri, Plate VI., fig. 10), is of a size intermediate between the two first of our collection, and the Twisted species. It grows in reddish tufts, being entwined with brown fibres, and having upright leaves lapping over each other. The urn is furnished with a very long lid. It is a rare Moss, and our specimens came from our kind friend at Blair-Athole.

There are a great number of these Screw Mosses; but those that we have here may serve as guides to the whole set. The Aloe-leaved species frequents clay banks, as does also the Bird's claw. The Slender Screw Moss, the Spreading-leaved Screw Moss, and the Hoary Screw Moss prefer chalk: while the Revolute and Convolute favour sandstone, and the Great Hairy Screw Moss grows on thatch. The rough-leaved and smaller Hairy Screw Mosses are parasites on trees.

Almost too nearly allied to the Screw mosses to constitute a separate group is the Water Screw moss family, the fringe has the same winding habit, but there is no stalk to the urn, or so short a one as hardly to deserve the name.

During our trip in the Highlands we had the pleasure of finding the smaller species (Cinclidotus Fontinaloides *Plate V., fig.* 11), in a rivulet near Oban. It was growing on a submerged stone, and the stems were borne along in the current. Some of them were five inches long.

The leaves are crowded and spreading, except when borne in one direction by the current; the urn is oval, and the fruit-stalk extremely short; the lid and the veil are both cone-shaped, the latter of a corky texture.

The Greater Water Screw moss is not found in Britain, except in a dwarfish, stunted form, growing high and dry.

The Screw mosses are succeeded by the Extinguisher

mosses (Eucalypteæ). They have oblong urns, beaked lids, and the veil is so large as to cover the whole urn like an extinguisher, hence the name.

Upon walls about Richmond we found the Common Extinguisher moss (E. Vulgaris) its branches were clustered, its urn lance-shaped, and the veil whole at the lower edge. The leaves were spreading and oblong.

The Fringed species was growing on walls on the road to the race-course, it has its name from the fringed margin of its veil (C. ciliaris).

Both these species have a fringe at the mouth of the urn, though that of the former soon vanishes.

The Sharp Extinguisher-moss (E. commutata) has no fringe ; and the Rib and Spiral fruited ones (E. rhabdocarpa and streptocarpa) have double fringes.

Between the extinguisher mosses and the grimmias there is a group of Alpine mosses, called Hedwigias, after a botanist of that name, or, in English, Beardless mosses, their urns being destitute of fringe.

The Hoary species (H. ciliata, *Plate V.*, *fig.* 12) I found above Callander, and afterwards among the Pentland Hills, it was growing on rocks, in large loose patches. The foliage was of a dark dull green and beset with hairy points, which gave it a hoary appearance. The urn was pale brown, and seated on the stem, and the mouth was very open, because the fruit was long past perfection.

The green Beardless moss (H. imberbe) is sometimes found growing with this, but is much more rare.

The large group of grimmias succeed these. They are tufted mosses growing on rocks, erect when little, prostrate when attaining any length. The lid is convex and slightly pointed, and the fringe single. The central column falls away with the lid : the veil is small. This group includes the grimmias and the Fringe mosses.

The close tufted and sessile grimmias (Schistidium Confertum and Apocarpum) have the base of the veil torn and jagged.

The Sea-side sessile grimmia (S. maritimum) Fanny found on slate rocks on the coast of Cornwall, its dense tufts of dull brownish green contrasting pleasantly with the golden and white Lichens. Its leaves are long and stiff, and its urn is seated on the top of the stem. Its fruit is perfect in the middle of winter.

The grey-cushioned grimmia (G. pulvinata, *Plate V., fig.* 14) grows abundantly on rocks and walls in most localities. It was thriving well upon the old bridge at Gunnerside. It is densely tufted.

Every leaf is terminated by a hair, which gives an appearance to the round cluster-like velvet pile; the fruit-stalks, which are erect now, are daintily arched in youth, so that the urn bends down again till it touches the leaves.

The round-fruited Grimmia has narrower leaves, and the Spiral Grimmia has striped urns. Schultz's Grimmia, and the Tall Alpine Grimmia, are much larger plants; and the Oval fruited, Hoary, and Dingy Grimmias have the base of the veil lobed. (*Plate V., fig.* 15).

The Fringe moss family (Racomitrium) resemble the larger Grimmias; they are tall and branched; the leaves are spreading, and often adorned with hairy points; and the veil is mitre-shaped and often cloven at the base; the urn is oval, the lid straight.

The Oval-fruited Fringe Moss is peculiar to the hills of Scotland, Wales and Ireland.

The Dark Mountain Fringe Moss grows on wet rocks by rivulets in similar situations; it has long stems, slightly clustered, and blunt leaves of a dull dark green. I found it near Callander, (R. aciculare, *Plate V., fig.* 18.)

The Slender mountain Fringe moss (R. sudeticum, *Plate V., fig.* 17.) was growing freely on rocks near Oban, its spreading leaves armed with hairy points giving it a grey appearance. It bore abundant fruit on short stalks, and had a venerable effect.

The Green mountain Fringe Moss (R. fasciculare) I found on damp rocks when making our memorable excursion among the Pentland Hills. The brighter hue, much forked branches, and absence of hairy points to the leaves distinguish the species.

The Hoary Fringe Moss is frequent in those Highland districts, spreading amongst heaps of stones and rubbish. Here the hairy points are very abundant, (R. canescens, *Plate V., fig.* 16), and the footstalk is longer.

The Woolly species (R. lanuginosum,) with its long slender brittle branches, and woolly pointed leaves, greeted us continually in our Swaledale rambles, forming thick mats large enough for the hall door, upon the loose walls on the hill sides. In some cases the stems were a foot long, with many branches, the foliage was a dull green, hoary with hairy points. The urns were of an oblong shape, and the fruit-stalks were short and rough.

These Fringe Mosses are a very handsome group, and

from their reverend gray hairs might be called the patriarchs of the family.

We carried our Screw Mosses, and Grimmias, and Fringe Moss home, treading softly over the carpet of Feather Moss in the low meadows by the river side, and thinking of the "green pastures and still waters" in which the Lord leads his people.

> "Praised be the mosses soft
> In the mountain pathways oft,
> And the thorns that make us think
> Of the thornless river's brink,
> Where the ransomed tread."

MOSSES. 79

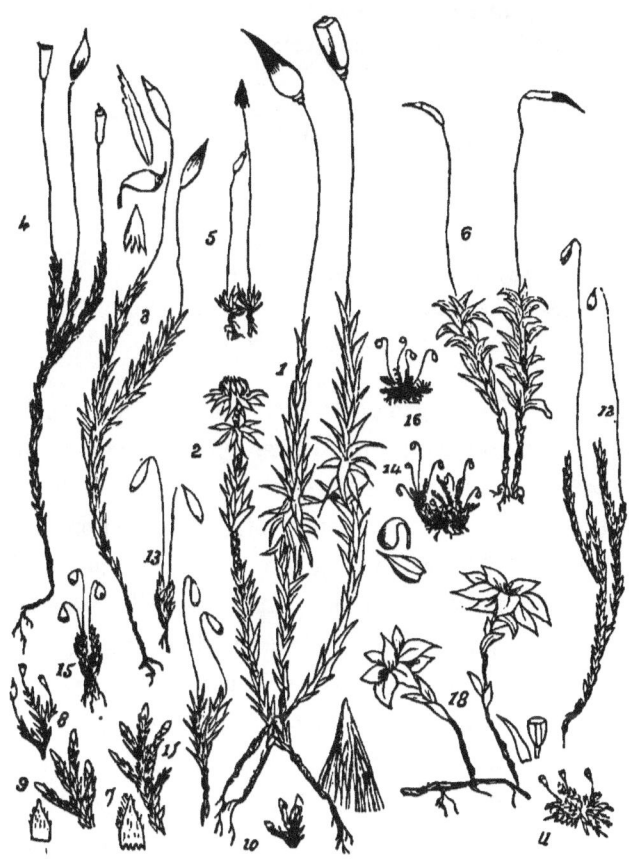

FIG. 1. Polytrichum Commune. 2. Secondary fructification of ditto. 3. P. Alpinum. 4. P. Urnigerum. 5. P. Aloides. 6. Atrichum Undulatum. 7. Orthotrichum Rupestre. 8. O. Crispum. 9. O. Affine. 10. O. Diaphanum. 11. Zygodon Viridissima. 12. Bryum Pallens. 13. B. Capillare. 14. B. Argenteum. 15. B. Cospiticium. 16. B. Carneum. 17. B. Cernum. 18. B. Roseum.

CHAPTER X.

> " Or to sit by the mossy fountain,
>   Where a sweet stream has its birth,
> And look around with admiring eye
>   On the lovely things of earth.
> The lichen, the moss, and the mountain-fern,
>   And the wild bee revelling there,
> And the bounding red deer, swift of foot,
>   And the bird that skims the air.
> For they link our souls to heaven,
>   And we feel the boundless love,
> And the wondrous power, and the matchless skill,
>   Of our Father who dwells above."

HAVING occasion to spend a little time in Richmond, we rambled pleasantly in the rich woods skirting the river, or climbed to the high moors, admiring the beautiful prospect, stretching panorama-like around us. One wood, rejoicing in the name of Billy Bank, was a favourite resort. An old wall separated it in one part from a meadow, and the droppings from the trees, and near vicinity of the stream, secured a perpetual damp, very favourable to the growth of mosses. We seat ourselves on this wall, collecting specimens of its verdant covering, and examining them with our pocket lens while yet in perfect freshness and beauty. Some members of the Bristle-moss group (Orthotrichum) are in our hands. The capsule is oval, and more or less embosomed in the leaves; the fringe is sometimes single, sometimes double, and in one species absent; the lid is short, cone-shaped, and beaked, and the veil bell-shaped, plaited, and often covered with hairs. The plants are in tufts, on trees or stones. The foliage is crisp and twisted when dry, but becomes spreading when immersed in water. There is always fruit upon the plant, for it takes twelve months to bring it to maturity. This Single-fruited Bristle moss seems quite at home upon the rough slab of limestone; its ripe urns have sixteen furrows, and the veil is wide and of a light colour. The Wood Bristle moss (Orthotrichum affine, *fig.* 9 *in cut*) adorns this old thorn stump, and luxuriates still more freely upon that pollard willow, and the Feather moss helps in clothing them. You might really suppose that Wordsworth had made his sketch here :—

> " Like rock or stone, it is overgrown
>   With lichens to the very top,
> And hung with heavy tufts of moss,
>   A melancholy crop :
> Up from the earth these mosses creep,
>   And this poor thorn they clasp it round,
> So close, you'd say that they were bent
> With plain and manifest intent
>   To drag it to the ground."

But on this occasion nature's poet is less tender to the plants than is his wont, and in the same degree less just. One can hardly look upon these dainty mosses as a "melancholy crop," nor imagine them entertaining a cruel intention towards the poor old thorn.

F

No : surely if mosses were able to give utterance to anything, it would be to a song of praise and love, setting forth God's care over all, even the very least of His works, and telling how He decks and brightens unavoidable decay by dressing the leafless tree with extraneous verdure, and how He smoothes the stony path, softening off the sharp edges by means of mossy cushions.

Let us now examine these other Bristle mosses. This with the erect branches, pear-shaped urn, and hairy veil, is the Rock Bristle moss (Orthotrichum rupestre, *fig.* 7 *in cut*); and this with the crowded crisp leaves and smaller urn is the Curled-leaved Bristle moss (Orthotrichum crispum, *fig.* 8 *in cut*). We must turn to the trees again for the Tawny-fruited Bristle moss, characterised by its dark foliage and narrow urn, and for the White-tipped Bristle moss, distinguished from all other species by the white points of its leaves (Orthotrichum diaphanum, *fig.* 10 *in cut*). There is a Straw-coloured Bristle moss, and a Close-tufted Bristle moss, and a Showy Bristle moss, and a River Bristle moss growing on trees by mountain streams; all these have furrowed urns and hairy veils. The Elegant Bristle moss has a smooth veil. Drummond's Bristle moss is peculiar to birch trees, and has creeping stems; and the Frizzled Bristle moss grows near the sea, and is always barren.

The group next to the Bristle mosses is a very small one, containing only one family and four species. The Yoke mosses have upright urns, small veils, and scarcely any beaks to the lids. They are distinguished from the Bristle mosses by the smoothness of the veils. I found a piece of the Green-tufted Yoke moss near Sheerwater, in

Wiltshire; it is rarely found in fruit, so I account my specimen a great treasure (Zygodon viridissimum, *fig.* 11 *in cut*). The Lapland and the Mougeot's Yoke mosses are Alpine species, and the Lesser Yoke moss is chiefly an inhabitant of Ireland. The Four-tooth mosses are two peculiar little plants. The Pellucid one (Tetraphis pellucida, *Plate VI., fig.* 11) has the secondary fructification in leaflets formed in a cup. Its leaves are broad and closely pressed to the stem, and its stems are matted together by rusty fibres. I found a quantity of it closely clustered between the decaying root of a tree and a mass of red sandstone rock in the Chase wood, near Ross. It grew like a miniature forest on a mountain side, and its full green hue and glossy lustre attracted my attention. Brown's Four-tooth moss (Tetradontium brownianum) has leaves of two kinds—the one broad and pressed to the stem, the other mingled with them, but narrow and spreading. Buxbaum's moss is entirely leafless, at least it appears so to the naked eye. It resembles a fungus, its stem being nearly buried in the earth, and only the reddish urn appearing above. The leafy Buxbaumia rewarded our perseverance in moss-hunting on one rather remarkable occasion. We were passing two or three days at Callander, and the last day was devoted to an excursion to the Leny Pass. It rained when we started, but having once agreed not to remark on the weather during our Highland trip, we only hoisted umbrellas in silence. On we trudged, admiring the beautiful glimpses we got of the mountain stream, and examining the wet banks in the hope of finding some of our favourite plants. A gentleman was bent on seeing the country as well as we, and he stood to admire at every point of beauty, but did not seek mosses. He was a foreigner, forming one of a large party at the principal hotel, where our "pursuit of botany under difficulties" formed a subject of mirth at that day's dinner-table. We saw and gloried in the grand rocks and waterfalls and precipitous woods; and on a low earth-topped wall we found nice plants, less than a quarter of an inch in height, and seeming like naked grains striped on the surface. These we carried home in triumph. The lens showed them like microscopic cabbages, surrounded by long, narrow, grass-shaped leaves, the urn shaped like a cone, and forming the heart of the cabbage (Diphyscium foliosum, *Plate V., fig.* 6).

Pursuing our path one day through the Billy Bank wood, we passed along a couple of fields, and came upon the hillock called the Round Howe. There the rocks and the river form a kind of circus, and in the midst of the area thus enclosed, the hill rises. The legend of the country is, that when the giants made the neighbouring hill, they had a spadeful of earth too much and tossed it down in this spot, so forming the Round Howe. Half of the hillock is wooded, and is an excellent place for wild flowers. But we did not linger there, being bent on finding mosses. In the thick wood which we next entered, the ground was carpeted with ferns, mosses, and lichens, and we hastened to gather specimens of the Wavy-leaved Hair Moss.

The large group of Hair Mosses is characterised by the spreading of the point of the central column which connects all the teeth of the fringe, but there are exceptions to this habit, where other points of agreement entitle the plants to a place in the group. The Moss we had just found had strap-shaped leaves, waved at the margin, and toothed: they were crowded on the upper part of the stem. The urn was cylinder shaped and curved, and the veils had long beaks. The old fruit had almost lost its shape, and the new was scarcely formed. Later in the year we got better specimens of the Wavy-Hair Moss (Atrichum undulatum, *fig.* 6.) It is a very common moss, flourishing freely in moist shady places on a clayey soil everywhere. The Hercynian Hair Moss is a Scotch and Welsh species (Olgotrichum hercynicum), it is of a firmer habit than the Wavy-leaved.

The Hair Mosses proper are very showy plants, woody in their stems, with rigid leaves, hairy veils, and secondary fructification in a star of leaflets. The name is given on account of the hairs on the veil. The Dwarf Hair Moss (Pogonatum nanum), we found on sod-topped walls in the Highlands. It is the smallest of the group with spreading leaves, and a roundish urn. The Aloe-leaved species (P. aloides, *fig.* 5), Edward found abundantly at Hawkhurst on the same clay banks where the Earth Moss flourished; here the leaves are lance-shaped, and the urn oval. When moist the leaves are spreading, but they cling close to the stem when dry. The Urn-fruited Hair Moss (P. urnigerum, *fig.* 4), was sent to us from the Westmoreland hills, and has a cylinder-shaped urn, its leaves are shorter, pointed and toothed, and of a glaucous tint. I knew that the Swaledale moors must produce members of this group, so I set out on an especial excursion to seek them, repairing once more to Summer Lodge Bank.

By the side of the hilly path, among Feather Mosses and grass plants, I found a tall Hair Moss with branched stems, narrow leaves and oval urns, its stem was bent at the base, as if it had thought of adopting a creeping habit. These characteristics proved it to be the Alpine Hair Moss (P. alpinum, *fig.* 3).

Passing by the verdant patches of swamp, and not pausing again to examine the thick white carpet of Bog moss, I came to where the ling, now in full bloom, was surrounded by a miniature pine forest, each lilliputian tree being crowned with a slender stalk a couple of inches in height, bearing a square urn at its summit. Many of these urns were naked, having lost both veil and lid, but a few belated plants still wore the veil, which, thickly covered with hairs, proved its wearer to be one of the Hair mosses. Surely, then, this was the Common Hair Moss (Polytrichum commune, *fig.* 1,) of which I had heard so much. Even in old Gerarde's time this was a familiar moss; for he says of it, "This kind of moss, called Muscas capillaris, or Golden Maiden-hair Moss, is seldom found but upon bogs or moorish places, and also in some shadowie ditches where the sun doth not come." This is the moss of which travellers speak as accompanying the Reindeer moss, and forming

1. HEDWIGIA.　2. GRIMMEA.
3. ORTHOTRICHUM.
4. POLYTRICHUM.

along with it the food of that useful animal, the two constituting the sole verdure of the snowy regions. I placed some specimens in my box, and proceeded on my way.

There is a Northern Hair-moss, (P. sexangulare, with leaves turning inward at the point, and six-sided urns; and a Slender Hair-moss (P. gracile), with densely tufted stems, and oval urn with indistinct angles; and a Buff fruited Hair-moss (P. formosum) scarcely at all branched, and with a thick buff urn. We did not find any specimens of these members of the family. The Bristle-leaved Hair moss, we found on hills in Shropshire (P. piliferium), it is characterised by sharp hair-like points at the end of the leaves, the urns are long and square. The Juniper-leaved species, with its slightly squared urns, and glaucous foliage, was sent us from Blair Athole, (P. juniperium.)

The Timmia (T. austriaca) is a rare moss, about an inch high, in form something like a Fox's brush; we have found no specimen.

When wandering one damp February day in the beautiful woods about Ross in Herefordshire, I came upon a block of red sandstone, forming a perfect nursery ground for flowerless plants; grotesque lichens were there in several varieties, and infant ferns, and delicate liverworts; but the plant most interesting to the subject under consideration, was the Bud-headed thread moss, which covered large areas of the rock with soft green crowded branches. What especially rivetted my attention was a quantity of tiny knobs, like minute rounded pin's heads, standing on short stems, all proceeding from the little moss. The curious heads (capituli) show little grains on the surface when examined with a lens, (Androgynum aulacomnion, Plate II., *fig.* 12.) There is a Marsh species (A. pausltre), resembling this, only larger.

The Slender Thread-moss (Orthodontium gracile) is a minute species, with short slender stems, branched and tufted. We have no specimen. Our kind Blair Athole contributor sent us the Golden Thread moss (Leptobryum pyriforme, Plate VI., *fig.* 13). It has silky thread-shaped leaves. Its pear-shaped urn, and fruit-stalk bent at the neck, show its relationship to the true Thread-mosses (Bryum).

These Bryums form a large group. They grow in tufts, on trees, rocks, and banks, with pear-shaped urns drooping elegantly, and leaves clasping the stem.

The Alpine bog Thread moss (Bryum pseudo-triquetrum), was sent to us from the Highlands. It belongs to a group of Bryums with ovate leaves, it has long stems, and grows in patches of a blackish green colour, the different styles of fructification are on different plants. From the same district came the Alpine Thread moss, (B. alpinum) its red tinted foliage and dense tuft making it remarkable; our friend could find no urns upon it. In our pretty Swaledale fern-wood we found patches of pale reddish moss on the margin of the stream, the stems were about an inch long, and thinly clothed with oval pointed leaves. The urn did not droop so much as in most of the species. It proved to be the pale-leaved Thread moss (B. palleus, *fig.* 12). We found it again in great quantities on the banks of the Swale about Richmond. The walls between Richmond and Easby nourished small forests of drooping urns. Here we found the Drooping Thread moss, (B. cernuusm, *fig.* 17,) with its fresh green oval pointed-leaves, and small urn ; the greater matted species (B. capillare, *fig.* 13), its leaves boasting a thickened border and a hairy point, the commonest of the genus ; and the lesser matted Thread moss (B. cœspiticum, *fig.* 15), with its rosy tinted urn. Near Hastings we once found a tiny Thread moss with deep red urns, it was the Bloody species, (B. sanguineum, *fig.* 16.) About Edinburgh I have seen garden walks and pavement covered with a small moss, its glossy foliage contrasting charmingly with the purple drooping urns ; the leaves are short, concave, and pointed, this is the dark purple thread moss, (B. atro-purpureum.) Growing on walls and by way-sides we find the Silvery Thread moss, (B. argenteum, *fig.* 14), growing in glittering patches, its urn is reddish, but rounder than that of the Purple species, to which the whole plant bears a close resemblance. We found it on the walls at Easby.

But the king of all the Thread mosses is the Rosaceous (Bryum roseum, *fig.* 18). The leaves grow in a cluster at the top of the stem ; they are broad and bend outwards, so as to resemble in form the petals of a rose. When first I saw it, by the side of the Twisted Screw moss, I imagined it to be a state of the Tunbridge fern, its leaves were so large, and its stems so long. It is rarely found in fruit. It is the largest species, its star-shaped cluster of leaves, adorning the ends of the branches procure its name. The stems are from one to four inches long. I first saw it in Goblin Combe, Somersetshire, and found it afterwards in a part of the ancient forest of Selwood in Wiltshire.

Fig. 1. Mnium Ligulatum. 2. M. Hornum. 3. M. Punctatum. 4. Paludella Squarrosa. 5. Funaria Hygrometrica. 6. Bartramidula Wilsoni. 7. Physcomitrium Pyriforme. 8. P. Fasciculare. 9. Bartramia Fontana. 10. B. Pomiformis. 11. B. Ithophylly. 12. Splachnum Ampullaceum. 13. Fissideus Bryoides. 14. Fissideus Taxifolius. 15. F. Adiantoides.

CHAPTER XI.

" From giant oaks that wave their branches dark,
To the dwarf moss that clings upon their bark,
What beaux and beauties crowd the gaudy groves,
And woo and win their vegetable loves ?"

GROUP of very handsome mosses nearly allied to the Bryums, succeeds that extensive family. They are distinguished from them in being called Thyme Thread mosses (mnium), they have generally large transparent leaves, and they make their new shoots from the lower part of the stem, not from the summit. With a view to collecting some of these, my friends directed me to a copse between the river and the road, near Gunnerside.

Under some bushes, upon a very moist bank, I had the delight of discovering a very large moss; its stems were tall, and surrounded with long strap-shaped transparent leaves, the vein along the centre being very pronounced; my pocket lens showed me that these leaves were jagged at the edge. Little branches grew from the tops of many of the stems in a cluster, and there were long creeping shoots also; a number of fruit-stalks rose from the summits of three or four of the plants, and the urns were still upon them, though the fruit had evidently been ripe months before. It was a much larger plant than any of the Thread mosses—even than the Rose Thread moss—

but its drooping urns at first inclined me to suppose that it must belong to that group; only, instead of being pear-shaped, they were oval. I hastened home with my trophy, and upon consulting my cousin's book, I found that the larger habit and oval urn distinguished the Thyme Thread moss group from that of the true Thread mosses. It was clear that my new specimen was the Long-leaved Thyme Thread moss (Mnium ligulatum, *fig.* 1 *in cut*). Marian congratulated me warmly on my success; for she said that this moss, though common enough, was rarely found with fruit upon it.

On our next excursion to Richmond, we spent some time in exploring the woods about Aske, the seat of the Earl of Zetland. A stream runs through the valley; and the thick foliage of the trees condense the vapours from it, thus forming an atmosphere most suitable for flower-less plants.

Here the mosses and liverworts were very luxuriant, and it needed no guide to draw my attention to a carpet of dark verdure under some holly bushes, which was all beset with drooping urns. The oval shape of these now over-ripe urns identified the plants as Thread mosses; the secondary fructification was also present in great abundance, forming plots of starry tufts; the leaves were broad and lance-shaped, and the fruit-stalk was very elegantly curved. It proved to be the Swan-neck Thyme-Thread moss (Mnium hornum, *fig.* 2). Another species, evidently belonging to the same family, was growing near; it had broad rounded leaves, which, when examined through the lens, were seen to be dotted, and edged with a thick border. The upper leaves were still broader than the lower, and arranged in a starry form ; the plants grew separate, not in clusters. The urns gave evidence of having been ripe early in the spring ; they were large and oval ; the leaves were of a very dark green. We at once agreed that it was the Dotted Thyme Thread moss (Mnium punctatum, *fig.* 3.)

In a wood behind my cousin's house in Swaledale, where rocks abound, and birch and willow trees droop over oozy places, I found another of these Thyme Thread mosses. It was covering the perpendicular surface of shady rocks, often spreading over their tops, too—its rounded leaves terminated in a sudden sharp point, procuring for it the name of Pointed Thyme Thread moss (M. cuspidatum). The stems were numerous, each bearing a footstalk and urn, the latter much inflated, egg shaped, and with a blunt lid. There were a number of long barren branches trailing on the rocks, giving the plant a likeness to a Liverwort. Here, too, I found the Long-beaked species (M. rostratum) resembling its neighbour in habit, but with broader leaves and more slender urns. The serrated species (M. serratum) I have got more recently from shady banks in the Herefordshire woods. Here the leaves are narrower than in either of the preceding mosses, and the lid of the urn larger and more pointed. The plants were growing in a scattered fashion. Our kind ally from Blair Athole has supplied us with the Round-fruited Thyme Thread moss (M. subglobossum) the denizen of marshes. Its leaves are large and broadly ovate, and the urn small and round. The many fruited species (M. affine) bearing two or more urns on a stem, I have found in woods in Swaledale. There is a Short-beaked, a Large-leaved, and a Star-leaved species, but they are all rare in Britain, and we have none of us met with them.

The Cupola moss (Cinclidium stygium) group succeeds that of the Thyme Thread moss; there is only one species, and it closely resembles the Dotted Thyme Thread moss, only the stems are matted together with purple threads, which give a sooty appearance to the plant, and its urns are pear-shaped. It grows in bogs.

A kind friend sent me a specimen of the Drooping-leaved Thread moss (Paludella squarrosa, *fig.* 4), which had been sent to her from Knutsford Moor, Cheshire. It had no fruit, she said, and the long stems were closely clustered together, the leaves being turned back. She had a drawing of the fruit upon a foreign specimen, of which she favoured me with a copy. The group of mosses called after Mees, the botanist, has only two British members. The Long-stalked Meesia grows scantily in Ireland, and the Dwarf Meesia is occasionally found among the Scotch and Welsh hills, along with the Lesser Pale Thread moss (Amblyodon dealbatus).

While rambling in the Richmond woods we came to a place where dead leaves and branches had been burned. A quantity of moss had sprung up among the ashes, resembling a Screw moss, only the fruit stalk was twisted in every direction, the urns were bent, and the dry ones furrowed; the veil was inflated below, and ended in an awl-shaped beak. The twisting of the fruit stalk shews it to be a Cord moss, and its upper leaves drawn close together, its red bordered lid, and notched border to the mouth, identified it as the Common Cord moss (Funaria hygrometrica, *fig.* 5). This moss always flourishes upon wood ashes, though it is obliging enough often to exist without them. I never saw it so abundant or so luxuriant as in Herefordshire, where they regularly level a portion of the extensive woods annually, using the trunks for timber or mine props, according to the size, and burning the refuse for charcoal. These old charcoal pits are like fairy gardens, carpeted with Cord moss and Liverwort, and hedged round by Arabis Thaliana, Myosotis Arvensis and several other pretty field plants.

There are two other British Cord mosses, but both very rare. The Irish Cord moss has been found near Cork; and Muhlenberg's Cord moss has no distinct British habitat.

The next family form a group called Bladder mosses (Physcomitrium) the characteristics of which are, a pear or club-shaped urn, upon an erect or slightly curved fruit-stalk, a convex lid, and an inflated veil. The bladder-like veil gives the name to the group. The Narrow-leaved Bladder moss has the urn erect, and the leaves lance-shaped and serrrated (Physcomitrium ericetorum). It inhabits heathy districts, and was sent to me from Teesdale.

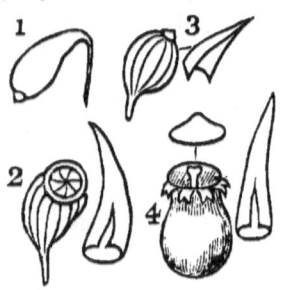

1. BRYUM.   2. FUNARIA.
3. BARTRAMIA.
4. SPHLACHNUM.

Edward found another Bladder moss. It was growing in patches, had oblong leaves, and a pear-shaped urn. It was the Common species (P. pyriforme, *fig.* 7). Another in the same district which he found in fallow fields about Hawkhurst, its fruit ripe in April, had narrower leaves and a tapering urn; it proved to be the Fallow-field Bladder moss (P. fasciculare, *fig.* 8). The Dwarf Bladder moss is very minute. We have none of us found it.

The Apple Moss group succeeds the Bladder moss; they are clustered mosses, growing upon rocks or earth; with roundish urns, ribbed when dry, a small cone-shaped lid, and a diminutive veil, which soon vanishes. In some species there is a simple fringe, in some the fringe is double, and in some it is entirely wanting. The Beardless Dwarf Apple moss is very lovely and delicate (Bartramidula Wilsoni): it grows in small patches, with decumbent stems, which are one-branched; the leaves are lance-shaped, and the fruit-stalks often grow two or three together, and are arched. The round urns droop, and are of a pinkish colour. The Rigid Apple moss, is peculiar to Ireland. This Common Apple moss, which we have from Richmond (Bartramia pomiformis, *fig.* 10), has clustered stems and long spreading leaves, which are crisped when dry.

Once more we planned an excursion to Summer Lodge Bank, but separated when we reached the mosses, that each might find different treasures.

For a long time I searched in vain for any indication of Apple mosses, but at last I found the object of my desire growing knee-deep in a rivulet which issues from the Tarn. The moss was tall, its stems measuring three or four inches, branched and matted together; the leaves were broad and tapering, of a yellow-green colour; and the long fruit-stalks bore large round reddish urns. It was undoubtedly the Fountain Apple Moss (Bartramia fontana, *fig.* 9).

Upon a bank crowned by rocks I found the Straight-leaved Apple Moss (Bartramia ithophylla, *fig.* 11). Its leaves are broadish at the bottom, but become very narrow and awl-shaped; they are of a light yellowish-green, and their clustered stems make pretty little cushions.

These were all the Apple mosses that Swaledale furnished us with. Subsequently I had the pleasure of finding two of them in the Highlands. Large elastic cushions of bright green, growing among rocks in the vicinity of Callander, the stems being two or three inches long, and the long slender leaves spreading in every direction, and bearing round urns on short upright stems, answered Hooker's description of Haller's Apple Moss (B. Halleriana, *Plate VI., fig.* 14). The still taller stems of the Curved-stalk species (B. arcuata, *Plate VI., fig.* 15) I found in wet places in the same district, but without fruit, here the branches grow in bundles, and the fruit-stalks are beautifully arched. We have since received specimens of this moss in full fructification, along with some of Oeder's Apple Moss (B. Oederi), of smaller size and darker hue, from our Blair-Athole friend. The Rigid species is very minute, with slender branches in bundles, and the urn large in comparison to the other parts of the plant, it is an Irish moss. The Thick-nerved Apple Moss (B. calearea) frequents limestone districts, its foliage is of a fresh green, and the long slender footstalk bears a very large urn. The leaves are broader than in most of the species.

The Cone Fringe Moss (Conostomum boreale) is allied to the Apple mosses by its round urn, but has a beaked lid and a longer veil than they. The leaves are lance-shaped, overlapping one another in five rows, and the stems grow in a dense cluster. It grows on the summit of mountains.

The Lurid Apple Moss (Catoscopium nigritum, Plate VI., fig. 16) has the fruit-stalk suddenly bent at the neck, it is two or three inches high, and grows in soft green tufts. The Naked Apple Moss (Discelium nudum) is distinguished by a large conical lid, and long awl-shaped veil.

But to return to our ramble on Summer Lodge Bank. The beauty of that September day tempted me to wander aimlessly hither and thither upon the moor. Now and then a distant gun bore evidence that the partridges had not entirely drawn away the foe from the grouse; and still as I roamed, the bird rose from among the ling and fled, uttering noisy cries. The hills basked in broad sunshine, across which cloud-shadows sailed like ships along a golden sea, and not a sound was heard from the valley, though a number of heavy wagons, laden with ore, were descending into it from the opposite hills. While thus lingering, my eyes suddenly fell on a clump of moss, thickly set with cylinder-shaped urns. Upon stooping to gather it, I found it was growing upon sheep-manure; the oval-pointed leaves were spreading, and the stems less than half an inch long. The fruit-stalk was very long, rather waved, and red; the urn was situated upon a large tubercle; the veil had fallen off. I now hastened home in good earnest, for I wanted to study the Collar moss group, to which I suspected that my new treasure belong. The tubercle decided the question, that being the leading feature of the Collar mosses.

According to Hooker, the veil in this group is small, cone-shaped, and torn at the base, and the lid convex. I could not judge of either of these particulars, for the fruit of my species was over-ripe, and the veil and the lid had perished (Splachnum sphæricum, *fig.* 12). There is a Large-fruited Collar moss, growing by springs in mountainous places, and a Flagon-fruited Collar moss, flourishing on manure in low situations.

Unfortunately, it was not a cavernous neighbourhood, so I had no chance of finding the Cavern moss (Schistostega osmundacea, *Plate VI., fig.* 17). This moss is of a pale glaucous green, very slender, and not reviving in water after it has been once dried. The urn is very small and oval; it has no fringe, and its tiny veil soon perishes. The delicate young shoots have often been taken for a Conferva; they have a refractive power, and on this account are said to illumine the gloomy caverns with a "golden green light." The plant is most frequently found on sandstone (*Plate VII., fig.* 1).

Several allies of the Collar mosses, as Tetrapladon, Tayloria, Dissodon, and Odipodium succeed them, but they are all rare Alpine species, and did not reward our search at that time or since.

The Flat-fork mosses (Fissidens) follow the Cavern moss in natural order. They are very small plants, with the leaves placed alternately on either side the stem, so as to be flat before and behind, something like minute ferns. The urn is oval, sometimes erect, but more often with its head a little bent; the lid is shaped like a mitre; but the most attractive part is the crimson fringe. I found one with fruit ripe in January last, in Kent. I did not then know its name or relations; but I was examining it afresh the other day: and by the thick border to its leaves, its fruit-stalks springing from the top of the stem, and its long lid, I decided that it must be the Common Flat-fork moss (Fissidens bryoides, *fig.* 13).

There are several other species with the fruit-stalk springing from the end of the stem: the Slender Flat-fork moss, the Green Flat-fork moss, and the Alpine Flat-fork moss. All these bear their fruit in winter. In the Fern-like Flat-fork moss (F. adiantoides, *fig.* 15) the fruit is borne half-way down the stem. It grows in those beautiful combes of Somersetshire, and I found it full of fruit last February. We frequently find the Yew-leaved Flat-fork moss, which bears its fruit on stalks springing from the sides of the branches. A number of stems rise from one base, and extend on every side in a half-procumbent manner; the urn is turned to one side, and, though you cannot see it thus late in the year, the lid is nearly as long as the capsule, and the veil is white (Fissidens taxifolius, *fig.* 14). This moss ripens its fruit in March. Nearly allied to this is the Marsh Flat-fork moss, whose branches creep to the extent of one or two inches; and the Short-leaved Flat-fork moss, which prefers fallow ground.

The chief interest that attaches to these delicate little mosses arises from the fact that Mungo Park gathered specimens of them in the interior of Africa, which specimens are in the hands of the authors of "Systematic

English Botany."*   It is supposed that he referred to one
of those plants when he relates that, in a moment of
despair, having abandoned himself to death, and believing
that the care of God's providence was no longer extended
over him, the extraordinary beauty of a very small moss
caught his eye.   He looked at its exquisite workmanship,
wondered at its adaptation to the barren home where it
was placed, and a train of softening thought swept over him,
analogous to the reasoning of the All-wise teacher: "If
God so clothe the grass of the field, how much more will
he clothe you, O ye of little faith?"   The voice of God
reached the traveller's heart by means of this diminutive
plant, and he arose armed with fresh courage, and went
on his way relying on his Lord.

* The specimens of Mungo Park are a variety of *Fissidens bryoides*
of Wilson (*Dicranum bryoides* of older botanists).

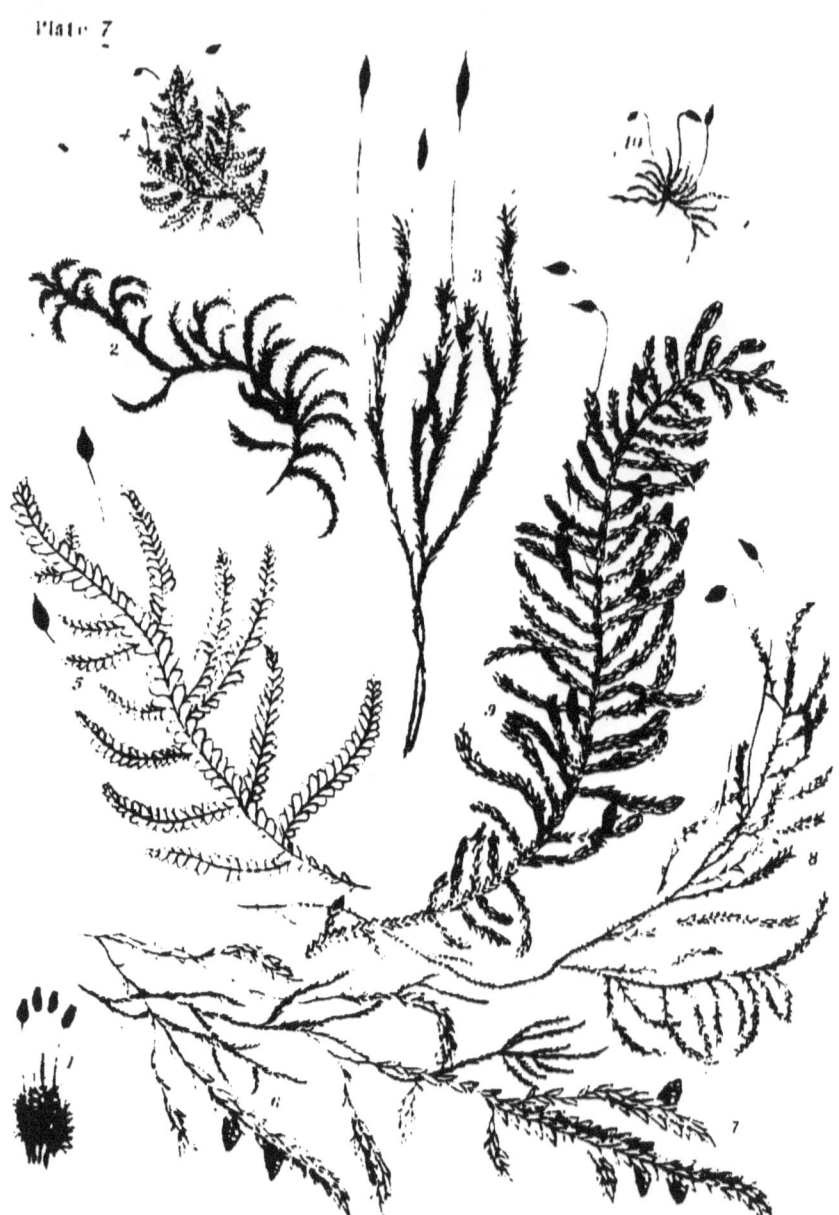

1 Tetraplodon. 2 Squirrel tail Leucodon. 3 Tall Anomodon. 4 Blunt Fern like Feather moss. 5 Flat leaved Neckera. 6 Greater Water moss. 7 Alpine Do. 8 Beaked Water Feather moss 9 Neat Do. 10 Neat Mountain Do.

CHAPTER XII.

> " And close behind this aged thorn,
>   There is a fresh and lovely sight,
> A beauteous heap, a hill of moss
>   Just half a foot in height.
> All lovely colours there you see,
>   All colours that were ever seen;
> And mossy network too is there,
> As if by hand of lady fair
>   The work had woven been."
>
> WORDSWORTH.

WE had now worked our way to the second great section of the moss group, the Side-fruited mosses. In this section the fruit-stalk springs from the side of the stem. Some small mosses, resembling the Beardless mosses, and characterized by the wide-mouthed urn, stand first in the Side-fruited group. The Compact Beardless moss (Anæctangium compactum), with its lance-shaped leaves and oval urns, is an Alpine species. Its brother, Hornsthub's Beardless moss, prefers Ireland as its native land.

Their allies, the Leucodons, grow on rocks or the bark of trees; the branches are incurved, and the leaves closely lapped over one another. The Squirrel-tail Leucodon (L. sciuroide, *Plate VII., fig.* 2) is often to be met with, but I have never found it in fruit. The English name is a good index to the plant. The Hare's-tail Leucodon (L. lagurus) is very rare, being only known to exist in the Hebrides.

The Wing mosses have curved fruit-stalks, are branched, and have oval urns. The Pendulous species (Antitrichea curtipendula), grows either on rocks or trees, three or more inches high, straggling in habit, and with hairy leaves. The Curled species (Leptodon smithii) is a Devonshire moss : its lid is beaked and the veil hairy.

The tall Anomodon (A. veticulosus, *Plate VII.*, *fig.* 3) is common at the roots of trees and on rocks. It has long branches, interlacing one another, so as to form a thick cushion. The leaves are tongue-shaped, and closely planted around the stem. When moist they are spread widely, but when dry they cling close to the stem. I found it with abundance of oval urns upon it, near Brixton Deveril, in Wilts.

One of the last days of my sojourn in this lovely Yorkshire dale was spent in seeing a lead mine and smelt mill, varieties with which these hills abound. As we left the harsh-looking neighbourhood of the mines, our eyes rested with delight upon a verdant plot of swampy ground. On inspecting this more closely, I found a quantity of yellowish moss, bearing clustered branches intermingled with fruit-stalks : the urns were erect and oval. The striking resemblance of its form to that of a tree rendered it probable that this was the Marsh-tree moss (Climacium dendroides, *fig.* 3), and all its characteristics agreed with Hooker's description.

The next time that my cousin went to Richmond, Fanny and I accompanied him ; and, alighting from the carriage a couple of miles before reaching the town, we followed a footpath leading through the woods by the river's side. In the lowest and most shady part of the wood we found the Fox-tail Frond moss (Isothecium alopecurum, *Plate VIII., fig.* 2), both carpeting the ground and hanging, tapestry-like, from the rocks; its short oval urns were already developed, though the weak young stalks were variously bent, and not as upright as they would be later in the season. This moss also has a tree-like habit, but it varies from the true Tree moss in being twice branched. Surely this must have been the suggester of the friendship between the moss and the poet in Dana's Plea :—

> " He praised my varied hues,—the green,
>   The silver hoar, the golden brown ;
>   Said lovelier hues were never seen ;
>   Then gently pressed my tender down.
>
> " And when I sent up little shoots
>   He called them trees in fond conceit ;
>   Like silly lovers in their suits,
>   He talked, his care awhile to cheat.
>
> " I said, I'd deck me in the dews,
>   Could I but chase away his care ;
>   And clothe me in a thousand hues,
>   To bring him joys that I might share."

The Leskeas are pretty feathery creeping mosses, with oval urns on erect stalks, distinguished from Hypnums, which they much resemble, by this peculiarity. The Many-flowered Leskea (L. polyantha) forms a pretty covering for rocks or tree stumps, bearing its myriad urns like ears of corn over the surface of a field. The Silky Leskea (L. sericea, *Plate VIII., fig.* 1) is a beautiful and common species. We found it abundantly at Easby. The branches are crowded along the stem, generally curved inwards; the lance-shaped, pointed leaves are pressed closely to the stem, lapping over one another; and the colour varies from deep olive green near to the root to sheeny gold at the tips of the branches. The Reddish Leskea is still more beautiful, its branches broader and beautifully tinged with red. Our specimen is from Blair Athole (L. rufescens).

The Irish Daltonia (D. splachnoides) is a minute moss, with a roundish urn, and beaked lid. Like the Leskeas, it has the fruit-stalk erect. The Lateral cryphæa (C. heteromalla), a rare Scotch moss, grows in a creeping manner, and has the urns seated on the branches without stalks. There is a larger variety which grows submersed in running water. Neither of these mosses grace our collection.

To procure the greater Water-moss (Fontinalis antipyretica, *Plate VII., fig.* 6), we sought the suitable habitat, a running stream. Being then at Richmond, we thought Skeeby Beck a likely place, and faced the tedium of a long walk on a dusty road to reach it. We were well rewarded for our pains. A large handsome moss attached to stones, its long stems borne forward by the current, its dark green leaves broad and pointed, and its urns seated on the stem half buried in sepal-like leaves, proved to be the species we sought. We drew handfuls of it from the stream, finding it beset with fresh-water mollusks and insects. Afterwards we found it in streams in Swaledale, wherever the water is not rendered unwholesome by lead-washings. Edward has it from ponds near Hawkhurst, and I have found it in the Wye near Ross; indeed it is common in running water all over the kingdom. Its brother, the Alpine Water-moss (F. squmosa, *Plate VII., fig.* 7), is a more slender plant, with glossy lance-shaped leaves, and urns like the common species. We have a good specimen from Blair Athole.

The Bristly Water-moss (Dichelyma capillaceum) is also a denizen of Alpine streams, with awl-shaped leaves ending in a thread-like point. It grows much in the same manner as the other floating mosses.

Repairing once more to Billy Bank wood, we found overhanging rocks clothed with luxuriant mosses. One had flat branches much divided, and lying one over another like tiles. The leaves were very small, olive green, and glossy, blunt at the ends. They were set on each side of the stems in a regular line, making a flat surface such as is called *complanate*. This proved to be the Blunt Fern-like Feather-moss (Omalia trichomanoides, *Plate VII., fig.* 4). In this genus the fruit-stalk is slightly bent at the junction with the urn, thus approaching the true Feather-mosses (Hypnum) in character.

Upon the same rocks a much larger moss was growing. Its characteristics were the same as those of the smaller species, but with erect urns. Its complanate foliage, pinnate stems, and tile-like growth, marked it as the Flat-leaved Neckera (N. complanata, *Plate VII., fig.* 5). There is a Crisped Neckera, with pale foliage, and a Dwarf

Neckera, growing on trees, but we have not found either.

The following spring, when searching those lovely Swaledale woods, we found a creeping moss, with large pale green glossy leaves, and stems rooting every few lines, and so tender that it resembled a liverwort rather than a moss. In due time the branches grew to an inch or two in length, and fruit-stalks rose bearing thick urns, with mitre-shaped veils (Hookeria luccus, *Plate VIII., fig. 9*). The deep green Hookeria is generally found in drooping caves in warm climates, as Ireland and Cornwall.

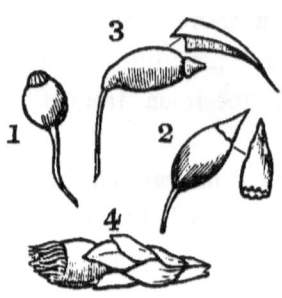

1. ANOMODON. 2. HOOKERIA.
3. HYPNUM.   4. FONTINALIS.

> " Drip, drip, drip,
>   In that cool and shady cave,
> From the basin in which the moss and fern
>   Their crumpled edges lave.
> Roofed by the living rock
>   That arches overhead,
> Ever by night, and ever by day,
>   Trickles that crystal thread.
> Ever in Summer's heat,
>   Ever in Winter's cold,
> Ever in Spring's young verdure,
>   Ever in Autumn's gold—
> Welling up from its secret urn,
> Purling its wreath of moss and fern,
> Pure and cool to the thirsty lip,—
> Ages have echoed that ceaseless drip."

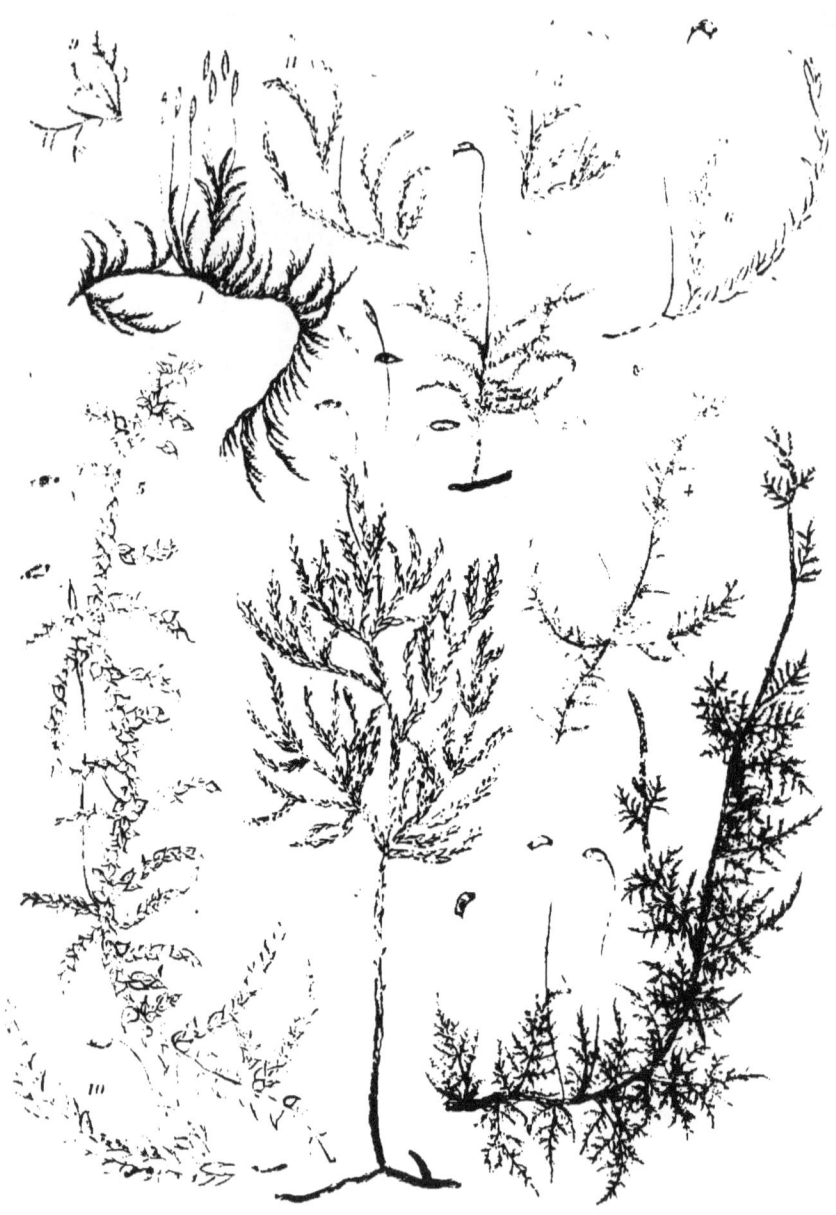

CHAPTER XIII.

MOSSES.

" Let long grass grow around the roots, to keep them
Moist, cool and green ; and shade the violets,
That they may bind the moss in leafy nets."
KEATS.

MY last ramble with my cousin was in a sub-alpine wood on the Spring End Farm, called Silk-wood. She guided me thither in the hope of finding Feather mosses, which, she said, were particularly beautiful there: all this extensive family have curved fruit-stalks, and the urns are generally slightly bent; their fringe is double; the veil splits on one side, the stems are creeping and tufted; the lid is cone-shaped, and most of the species ripen their fruit in the winter, or early in the spring. Fanny produced a specimen of the Rusty Feather moss (Hypnum plumosum) which had been sent to her from Teesdale: its stems are branched, and the branches elongated. The Rough-stalked Feather moss she had found in Wiltshire (H. rutabulum); it is a showy moss, with procumbent stems, arched, and often rooting at the extremity; the leaves are pale green, concave, and pointed; and the fruit-stalk is very rough.

We came upon a lovely cushion formed of Swartz's Feather moss, and the prolonged Feather moss; they were not in fruit, but their feather-like branches interlacing one another made a beautiful carpet. The common Polybody thrust its fronds through this cushion, and luxuriated in its protection for its creeping roots. It reminded me of the old legend of the ferns and mosses being at warfare, when

> "The fern loved the mountain, the moss loved the moor,
> For the ferns were the rich, and the mosses the poor."

At this period each kept to its own locality, and the sun scorched the mosses, and dried the roots of the ferns, while the wind beat pitilessly upon them; but affliction brought both to their senses, and they agreed to help one another; so the tall ferns shielded the mosses from the sun, and the mosses protected the roots of the ferns from the wind, and kept them moist. The Striated Feather moss (H. striatum, *Plate VIII.*, *fig.* 4), resembles these. It is a very common moss, covering tree roots and stumps, and twining among short grass, giving a downy mat to every rough woodland spot. I have found it in Wiltshire, Kent, Yorkshire, and Herefordshire, bearing abundant urns in the winter months.

We found the Wall Feather Moss (H. murale) growing on some rough stones in an exposed part of the wood; it is a delicate little moss, with short roundish branches; its leaves are broad and pointed, and the patches of it were pale green and shining.

A still more delicate species, the Creeping Feather moss (H. serpens), grew on that same heap of stone; it has tiny spreading leaves and oblong curved urns. Our attention was next arrested by a cushion of large branched moss, every leaf of which seemed as clean, and bright, and glossy, as if newly washed; the erect stems were simply branched, the branches curved slightly, and the roundish, closely overlapping leaves so arranged as to make the branches round, while the turned back points of the leaves gave them a bristly appearance; the urns were oval, and placed at right angles with the foot-stalk. There was such a puritanical appearance of propriety about the whole plant, or rather congregation of plants, that I rejoiced to know its name to be the Neat Feather moss (H. purum, *Plate VII., fig.* 9). Although it is a very common moss, it is rarely found in fruit, so that we justly regarded it as a treasure.

From early childhood I have loved one moss in particular, and I believe that my taste in this respect is shared by very many, both of old and young. I loved the Tamarisk Feather moss (H. tamariscinum, *Plate VIII., fig.* 3), while yet wholly ignorant of the nature of any plant; when I could not better describe my preference than by saying, "There is one moss I like, and one I don't like." The varying colour of its foliage, shading from yellow to myrtle green: the rich luxuriance of its branches, laid one over another like the feathers on the breast of a bird; the delicate arrangement of its myriad of tiny leaves, won my love then, and keep it still, when hundreds of moss beauties might divide my allegiance.

I fancy that this moss, or some member of its family closely resembling it, is the one of which Gerarde speaks as the "Mosse ferne." He says: "There is likewise found in the shadowie places of high mountains, and at the foote of old and rotten trees, a certain kind of mosse, in face and show not unlike to that kind of oke (oak) ferne, called Dryopteris. It creepeth upon the ground, having divers long branches, consisting of many small leaves, every particular leaf made up of sundry little leaves, set upon a middle rib, one opposite to the other." This moss is very frequent in our woods, but it is rare in fruit. I have found it with urns near Ross, and beautiful specimens from the neighbourhood of Carlisle.

The Triangular-leaved Feather moss was there too (H. triquetrum, *Plate VIII., fig.* 5), it is familiar to most people as the moss generally used in packing. You buy it in London dyed a coarse colour; and you see it fastened to paper flowers to represent the calyx leaves. It is a handsome moss; its stems grow five or six inches high, and its triangular leaves turn back; this moss also is but rarely found in fruit; our Silkwood plants, however, bore abundant urns.

Closely allied to this, and nearly as abundant, is the Rambling Mountain Feather moss (H. loreum), it is less rigid in its habit, and more slender, its branches are pinnate, the branchlets drooping, and the leaves are smaller and more crowded. It is also frequent in the Swaledale woods.

We were about to leave the wood, when I observed a mat of what I imagined to be a dead moss; the leaves were arranged on each side the stem, so as to present a flat surface; they were large and oval, and much waved and crisped; in colour it resembled a Bog moss. A great number of chestnut-coloured urns garnished the spare interlacing branches; these were dry, having been ripe in May, and were furrowed. Marian assured me that the whitish colour was as much a mark of the Waved

Feather moss (H. undulatum, *Plate VIII.*, *fig.* 6), as the form of the leaves, or the stripes on the urn.

There are a vast number of species in this over-grown family, nearly a hundred of them, I believe ; but Silkwood did not offer any more of them, and we had not time to search further.

Returning by Haverdale wood, we followed the course of the pretty stream for some little distance. The water is there drawn off into a mill race, and in this we noticed moss branches waving in the current. We got a handful, and were charmed to recognise the Long-beaked water Feather moss (R. ruscifolium, *Plate VIII.*, *fig.* 8). It seemed to be the very scene described by Gardner in his poem on that plant.

> "There you may look beneath the waters
> Sweetly gliding, or serene,
> For one of beauty's lovely daughters,
> Lovely, though of humble mein ;
> And where the stream in childish glee
> Leaps o'er the rocks in infant pride
> This little moss, in eddying swirl
> Of foamy waves, its head doth hide."

The mill race was only interesting as the home of the moss, and we turned from it to the brook-side, which presently became wild and beautiful in the extreme. The waters were now descending towards the broad valley by many an abrupt stony step ; then suddenly they fall into a circus formed of huge limestone rocks, making "eddying swirls" and "fancy waves" in abundance, and nourishing curtains of rich verdure on either hand. Here was the same moss again, and with urns upon it! The stems were several inches long, and irregularly branched, and the foliage was dark coloured, oval and somewhat complanate. The urns were thick, and had lids with long beaks. Upon the wet rocks, watered by the spray from the cascade, another Hyperum was growing and displaying its urns, which had also beaked lids. The leaves were spreading, the stems several inches long, beset with numerous slender branches; it was the Curled Feather moss (H. commutatum). Afterwards we found it on dripping rocks in the Clink Bank wood near Richmond, and there it was coated with lime which filtered from the rock. In the higher part of the wood was a wide mat of a branched moss, evidently a Feather moss. The stems were four or five inches long, twice pinnate, and closely beset with oval pointed glossy leaves, it was the Shining Feather moss (H. splendens), and had urns upon it, oval, and with a beaked lid. It is not common in fruit.

The Plumy Crested Feather moss I found afterwards in a thicket on the border of the Wiltshire Downs. The leaves are heart shaped, ending in long points which curl round, the stems are pinnate, the branches tapering elegantly. It is remarkably soft, a velvety looking moss, its foliage a yellow green (H. molluscum).

We have the Scorpion Feather moss (H. scorpioides). with its long stems, but slightly branched, its lurid twisted foliage and bent urn; the Neat Mountain Feather moss (H. pulchella, *Plate VII., fig.* 9), with its minute fresh green branchlets, and nearly erect urn; the Hook-leaved species (H. hamulosum) yellow in tint, and branched and clusters; and the Ostrich Plume Feather moss (H. crista castrensis) with its truly feather like fronds; from Blair Athole.

There are a great number of species in this genus, but these serve for examples: new specimens will give zest to new explorations.

It is one of the great advantages in this group of plants that they are in perfection when flowers and ferns have retired for the winter. As long as the weather remains open, mosses increase in beauty; and after a snow storm and many a cruel frost, the relenting breezes and soft rains of February soon succeed in reviving them. All moss collectors should preserve several specimens of each species which they are fortunate enough to find, as, now that the study of these plants is so greatly on the increase, they are sure ere long to meet other collectors who are very glad to accept or to exchange.

No wonder that the study of mosses is becoming popular, for it is an insensible mind indeed which has not taken pleasure in them from childhood. To step upon moss-covered ground sends a thrill of delight to many a heart, the attention of which has never been given for a moment to Monocotyledons, or Dicotyledons, or Acotyledons; and who is there that cannot call to mind the refreshment of throwing themselves down upon a moss bank such as Wordsworth describes?—

> "Here, traveller, rest thee, for the sun is high,
> And thou art old and weary. It is sweet
> To find at noon a moorland bank like this,
> To press its luxury of moss, and bid
> The hours fleet by on burning wing."

I remember finding a famous mathematician standing rapt in delight while beholding a group of rocks covered with moss and lichens, and I have often seen little children dance for joy at sight of a mossy hillock.

> " Why should greenness charm the eye ?
> Such is God's good will!"

Surely, while we admire this exquisite portion of God's handiwork, our hearts must break forth in the holy song : "Oh, all ye green things upon earth, bless ye the Lord ; praise him, and magnify him for ever."

MOSS ALLIES. 117

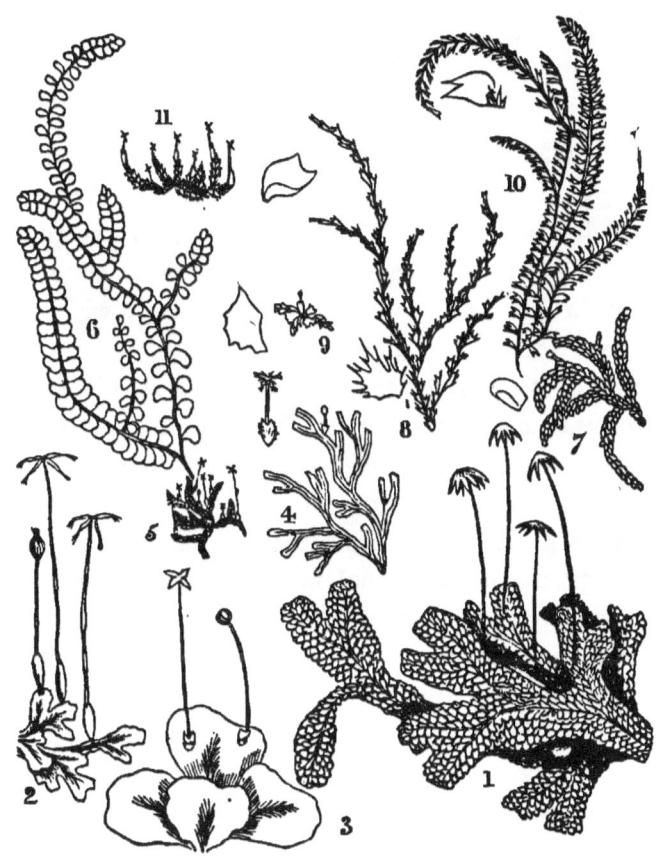

FIG. 1. Variable Marchantia. 2. Earth Liverwort. 3. Leaflike L. 4. Forked L. 5. Thyme-leaved L. 6. Asplenium-like L. 7. Complanate L. 8. Fringed L. 9. Small Downy. 10. Doubled-toothed L. 11. Inflated L.

CHAPTER XIV.

MOSS ALLIES.

"Ah life, I breathe thee in the breeze,
I feel thee bounding in my veins,
I see thee in these stretching trees,
These flowers, these still rocks' mossy stains."

BRYANT.

THERE is a group of plants closely connected with the mosses, and often mistaken for them, called Liverworts (Hepaticæ). These plants have either leafy branches or fronds, seeds affixed to spiral threads, and are propagated by buds, as well as by the seed. One large family and four small ones are contained in this Hepaticæ order. The true Liverworts (Jungermanniæ) bear their seed in round or oval capsules, which burst into four valves when the seed is ripe, the empty capsules remaining like brown stars or crosses upon the plant. These capsules are situated upon longer or shorter footstalks, of such fair and delicate structure, as closely to resemble spun glass. During the early stage of formation, the capsule is contained in a transparent veil, or wrapper, situated on the stem, and surrounded by calyx leaves; this veil opens after awhile, and the round head rises from it, the footstalk lengthening day by day, till the valves open and the seeds are emitted.

The Liverworts are divided into two great families:—
1st, Foliaceous Liverworts ; 2d, Frondose Liverworts.
The leafy or foliaceous Liverworts are again divided into
those without stipules and those with stipules: the stipules
are rudimentary leaves, situated between the true leaves,
on the under part of the stem. The first time that my
attention was attracted to the Liverworts, I was wandering in Longleat Park and its adjacent woods. It was
early spring, moist, warm, February weather. The
mosses lay thick and verdant at my feet—the Swan-necked Thyme Thread moss with its forest of half-formed
urns, and the Tamarisk Feather moss with its luxuriant
plume-like foliage. Among these entwined long delicate
branches, with roundish transparent leaves in exact rows
on either side ; the stems were brittle, and I found it
needed great care to disentangle them from the companion moss without snapping them off in short pieces ; the
leaves were pale green and shining, and a pocket lens
showed them to be formed of a net-work of cells ; and
there was no mid-vein. I felt sure that this could not
be a moss, still less a fern ; but as I could find no fruit,
I had to wait to determine its order. It was not till
afterwards that I ascertained it to be the Asplenium-like
Liverwort (J. asplenoides, *cut* 4, *fig.* 7), the largest and
commonest of the Leafy Jungermanniæ. In those same
woods, but later in the season, I found a similar plant
studded with capsules, some entire and puffed out with
the ripening seed, and others already empty, and standing
star-like on the transparent footstalks. I then recognised
in both the characteristics of the Liverworts.

Winter and early spring is the time for finding Liverworts—their delicate structure shrivels up before the heat of summer, and the growing plants and grasses hide them from sight before the heat makes them invisible. In the Chase Wood, near Ross, these plants flourish in perfection, luxuriating alike on the red sandstone rocks, the moist banks, or the tree roots. In some parts of the wood the ground is so steep, that one is glad to stop and rest before climbing higher, and then to examine the treasures of the rocks at one's elbow is a natural recreation. While thus engaged, my eye fell on a number of extremely slender dark green stems. I gathered some, and applied the lens. My first impression was that it was a land Alga, but minute leafy prominences on either side the stems were rendered evident by the lens. The microscope showed them to be heart-shaped leaves entirely clasping the stem with their broad leaves, and clinging so close to it as almost to seem one with it. It was the Heart-leaved liverwort (J. cordifolia).

Other rocks offered a great variety of minute flowerless plants for inspection, the crop being so abundant that but little of the warm colouring of the rock could be seen, beyond here and there a quartz pebble cropping out of the red sand, like plums from a cake. The Asplenium-like species was there in its glory, embracing mosses and infant ferns; and under shelter of those comparatively large plants, tiny Lichens like minikin pins, and Liverworts no taller than velvet pile, grew in microscopic beauty. My lens revealed starry capsules and glassy threads upon the green velvet, so my knife was put in requisition to procure a minute sod of the plant. With the aid of a small microscope, I found my Liverwort fully branched, and leafed, each leaf shaped like a perfect crescent, it was the crescent leafed species (J. connivens). A wood in Swaledale, already familiar as the scene of many a fern and moss-seeking ramble, furnished also its quota of Liverworts. Here the Asplenium-like species was rambling as was its wont; while on the deeply shaded rocks the closely pac ed erect stems of another species were forming a verdant cushion. The leaves seemed to be placed in four rows, but on closer examination it appeared that there were really only two rows of leaves, but that each leaf had two lobes, the smaller turned to the front of the stem; the leaves were of a pale green, and showed toothed edges under the lens; the calyx, veil, and capsula were placed on the summits of the branches (J. umbrosa). Another of the group was there, with leaves of similar form, but plain at the edges; it grew upon a barer rock, and its stems were shorter and unbranched; it was the Bluntleaved Liverwort (J. obtusifolia.) Upon beech boles in that Chase wood, were patches of green verdure. The plant stuck so close to the bark that you might have supposed it a mere painting. Here again the leaves were lobed, but the smaller lobes were turned in, they were planted close on the stem, so as to overlap one another. The footstalks were situated at the end of the stems, and were very short indeed, the capsule seeming as if seated on the veil. This was the Complanate Liverwort (J. complanata, *fig.* 9). All these belong to the group without stipules, many more belong to the same group, but we have not found them, and all these will serve as examples.

Later, in that same Herefordshire wood, the Polyanthus Liverwort bore an abundant crop of capsules (J. polyanthus). Here the foliage is very pale and transparent, the leaves are roundish, and placed close to one another on either side the stem, which is rooted at every few lines to the moist bank. Under the stem, between each pair of leaves, is a fork-shaped stipule, indicating its position in the stipule group. The Double-toothed liverwort (J. bidentata, *fig.* 10) is nearly as common as the Asplenium-like species. Its foliage is transparent like the one just described, and its stems soft and brittle, but its branches grow to a length of several inches, interlacing among moss, and its leaves are cut into two teeth. The veil and calyx are large, and the footstalks long and showy. We have found it in Wiltshire, Yorkshire, Kent, and Herefordshire. The same rock which nourished the Crescent Liverwort, gave a home also to the Creeping species (J. reptans). To the naked eye this looks like a starry cluster, delicately branched, but the lens reveals every tiny leaf to be cut into three points, and arranged side by side along the pinnate branches. Here, also, the capsules were abundant.

Above, upon the rock, hung large branches, with crumpled leaves, overlapping one another, two-lobed, and the smaller lobes turned inwards, round stipules occupying the space between the lobes, and making the branches look like green chenille. These branches were placed one over another like tiles, and gave refuge to numerous spiders, wood-lice, and sparrow-shells. This species is a very common one (J. platyphylla). Among scattered trees bordering one of our Yorkshire moors, upon ground oozy to the tread, we gathered the fringed Liverwort (J. ciliaris, *fig.* 8), so-called because of the hair-like segments into which the leaf was cut. The verdure of the mossy cushion attracted me, and it was not till I had it in my hand that we saw it was a liverwort. There, also, forming soft green cushions at the tree roots, was the Thyme leaved liverwort (J. serpyllifolia, *fig.* 5), its leaves placed in two even rows, and its general habit resembling the Asplenium-like species, but much smaller, and unmistakeably distinguished by the small indented stipules under the stem. The neighbourhood of Hawkhurst, in Kent, first furnished us with the Spreading Liverwort (J. dilatata); Edward found it growing in crimson patches on the bark of trees, and could only secure specimens by taking off a slice of the bark. It resembles the Complanate species in its broad overlapping leaves, but is distinguished by its toothed stipules. The colour of the foliage is not a sufficient distinction, for that of J. complanata is sometimes tinged with crimson. Upon clay banks in that district he also found two minute species, the Downy (J. pusilla, *fig.* 9), distinguished by its very large veils, and bluntly toothed foliage; and the Inflated (J. inflata, *fig.* 11) with short branches, and doubly toothed leaves. Neither of these have stipules, so they belong to the former group.

The Earth Liverwort (J. pinguis, *fig.* 2) stands first in the Frondosa family. We found it growing freely in the Aske Woods and on the Clink Bank, near Richmond. The fronds are cut into lobes, often shaded with purple towards the centre. The leaf-like Liverwort (J. epiphylla, *fig.* 3) is very common, growing on damp garden-walks and pots, as well as in moist woods and lanes. I have gathered it with its round capsules in abundance in Yorkshire, Kent, and Herefordshire. The Many-fingered species (J. multifida) is also found in the Chase wood along with the preceding one. It is distinguished by having the frond deeply and frequently cut. Billy Bank wood furnished us our first specimens of the Forked and Downy Liverworts (J. furcata and pubescens, *fig.* 4). In general form they are nearly alike—their fronds strap-shaped and branched, their capsules small, standing on a short thick footstalk, and proceeding from a hairy veil. The principal difference is in the one species being smooth, and the other covered with down. They grow in dense masses at the roots of trees, the fronds being then nearly upright and interlacing; or upon the bole of trees, when the fronds adhere closely to the bark, and the whole plant looks like a silky covering.

Such is the natural arrangement of the true liverworts, simple and easy of observation, and attractive from the exquisite fineness and delicacy of the members.

Another family in the Hepatica order is that of the Merchantia. Here the foliage is frondose, and palpably cellular, the fronds attaching themselves by roots to the ground. The fructification is contained in vessels placed under a round covering, which is raised on a long footstalk, and cut into numerous segments. Children call these heads of fructification, "little umbrellas." The Variable Marchantia I first found in fruit, by the side of a mill-dam at Vallis in Somersetshire, but I have since seen it in much greater beauty in the charcoal pits in Herefordshire. The year after the burning of the wood small plants of this Marchantia appear upon the ashes, and put up a number of small heads, on short footstalks; the next year the fronds cover the ashes, and a forest of stems rise, twice the length of those of the former year. The head becomes quite flat before the seeds ripen, and the numerous segments are connected for part of their length by a thin membrane. The Bell Marchantia (M. hemispherica) does not spread so widely, its footstalks are shorter, and the heads are bell-shaped, and cut into fewer segments. Our specimens were sent to us from the New Forest, Hants. The Cone Marchantia (M. conica) I gathered on Rudd Heath in Cheshire, its fronds were tipped with crimson, and its heads were cone-shaped. There are some curious plants included in the Hepatica order, which are very rare, and very unattractive.

The Riccias are floating plants, with minute fronds, fringed or lobed; they are found in ponds among Duckweed.

The Anthoceros has somewhat the habit of a cup-lichen, but bears its spores in a kind of sheath, a continuation of the footstalk; its habitat is clay banks.

The Targonia has an expanded frond cleaving to the earth, and having the seeds in a round or oval sac, situated on the margin of the lobes.

The Nardoo belongs to this order. Its foliage is divided into leaflets, and the spore cases are saucer-shaped, and situated upon long footstalks. It grows upon ground that has been inundated, ripens its seeds and sheds them, and when the floods return, they germinate and secure a fresh crop for the succeeding summer. These capsules occur in such abundance that they furnish an article of food to the inhabitants; Macpherson and Lyons lived for some time upon a pint of Nardoo a day. It is prepared by pounding and then making into dough cakes. The plants of this group are not known in Europe.

RAMBLES
IN
Search of Flowerless Plants.

CHAPTER XV.

SEAWEEDS.

" To the shores ! where the bright green sea
    Its snowy spray is throwing,
Down by the mystic looking caves,
    Where healthful winds are blowing ;
There cull the treasures of the deep,
    Where gems of pearly beauty lie,
Where sea-birds their carousals keep,
    Chiding the stranger wandering by."

FOR a long while we had talked of a tour in the Highlands, and two summers had passed by without our achieving it. A third August was far advanced before our various engagements left us free, and all our friends ridiculed the idea of a botanical tour in September. But as all was now ready for a start, my young cousins and I were equally anxious to put our plans in execution, so we agreed to seek our harvest at Oban and Arran, instead of among the hills, and to gather sea-weeds in place of flowers.

With this end in view we gave ourselves up to the enjoyment of the beautiful scenery along the Lakes Katrine and Lomond, and through the magnificent Pass of Glencoe, and we only began the business of our tour when we were quietly ensconced in lodgings at Oban.

It was low tide when we took our first ramble on the shore, the sea was as still as an inland lake, which indeed it there very closely resembles, the two horns of the bay seeming also to touch the extreme points of the long island of Kerrera which lies opposite to the town. Heaps of waste of every shade of olive lay upon the sands, and

> "As we strolled along
> It was our occupation to observe
> Such objects as the waves had tossed ashore,
> Tangle, or weeds of various hues and forms,
> Each on the other heaped, along the line
> Of the dry wrack."

The curled fronds of the Tangle which fringed the margin of the water rose and fell lazily as the slight motion of the waves stirred them, and the Star-fish lay at ease among the stones to which the weeds were attached.

I pointed to the heaps of rubbish. "There lies our first lesson," I said; "let us sit down on the shingle, and carefully examine each weed in that black ridge.

The whole mass was of an olive colour, though the branches that composed it were of varied form and shade. By this I knew that all the weeds belonged to the first great division, one of the characteristics of which is the olive colour. In the Algæ colour forms a permanent distinction, and the whole family, including both the denizens of the sea and of fresh water, are ranged in three divisions : the first *olive* in colour, with seeds or *spores* also olive, and a secondary form of fructification called *antheridia*, containing orange bodies, which, when seen under the microscope, exhibit lively motion as if of animal life. The second division are *red* weeds. They have fruit of two kinds, each one on different plants; the one called *spores* is contained in cases dispersed through the substances of the weed ; the other, called tetraspores, is often external. The third division contains, as a rule, *green* weeds, though there are some exceptions in favour of purple. The *spores* are often endowed with motion, and the *vesicles*, or secondary form of fruit, are external. Some of these green weeds belong to the sea, but many inhabit freshwater. In size, Algæ vary from simple microscopic cells to branched woody plants, many fathoms in length. The structure of them all is very simple, consisting of roundish cells either adhering firmly to each other, or connected by transparent gelatine.

The first heap we examined was in a great measure composed of the blackish strap-shaped fronds of the Sea-oak. These were branched and thickened at intervals of less than an inch with bladder-like air vessels. The whole plant looked like strips of thick dark leather. This is the first British member of the Olive family, or *Melanospermæ*.

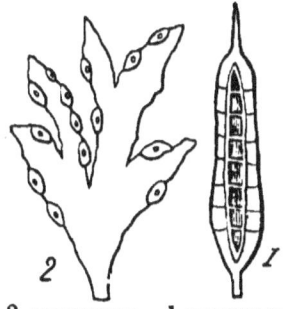

2 SARGASSUM. 1 HALIDRYS.

Its botanical name is Halidrys siliquosa (*Plate I.*, *fig.* 1).

The Gulf-weeds (Sargassum, *fig.* 8) precede the Sea-oak in order; but although they are occasionally washed upon our shores, they are not really natives. They are well known to sailors as floating in enormous masses in the North Atlantic Ocean. Columbus was in despair when he encountered one of these banks of weed, and seeing the ship so seriously obstructed by it, he believed for a time that God had frustrated his undertaking. The Gulf-weed is branched and leafy, and beset with stalked air vessels which look like berries.

Edward described a weed which he had found in abundance on the coast of Cornwall. It had air-vessels like berries, only they occurred in the substances of the stalk, which was thus made to resemble strings of beads, and little branches springing from the side of these vessels, and bearing narrow flat spines, not unlike juniper leaves in form. The disposition of the air-vessels proved it to belong to the family of Cystoseira, the name of which is taken from two Greek words, signifying *bladder* and *chain*; and the flat-pointed pinnules or *fibres* distinguish it as the species *Fibrosa* (*Plate I.*, *fig.* 2). Harvey states that this species is not found in Scotland. Indeed, it is rare to meet with any Cystoseira there. We were lucky, therefore, to have a specimen from Cornwall for our collection.

There is a Heath-like Cystoseira found on the south coast of England, which bears a great resemblance to the plant from which it is named—a Granulated species and a Fennel-like species.

A specimen of the Pycnophycus tuberculatus, the only British member of the fourth family in the Fucus order, was sent to me from Jersey. It has a rounded forked frond, and is very brittle, as my fractured morsel proved.

But to return to the heap at our feet, where fair specimens of several of the true Fuci were lying: there were large forked fronds of a dark colour, and with sharp cuttings on its margin like the teeth of a saw: that was number 1, *Fucus serratus* (Plate I., *fig.* 3), Then there was a great branch with large air-vessels and thick flat branches like a coarse cystoseira; that was number 2, *Fucus nodosus*. A smaller plant 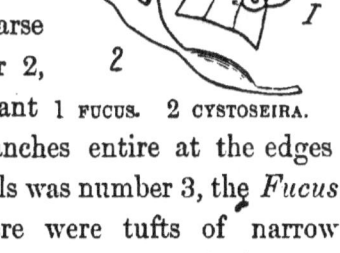 1 FUCUS. 2 CYSTOSEIRA. of paler olive, with forked branches entire at the edges the tips swollen with air-vessels was number 3, the *Fucus vesiculosus;* and finally, there were tufts of narrow forked fronds, channelled in the centre, and this was number 4, the *Fucus canaliculatus.*

Here was certainly an excellent start for our collection, and these unattractive weeds are really the most important in their uses. In olden times, "vile Algæ" was a term for anything utterly useless, but this error has long since died away. Once on a day, some country-folk gathered a bundle of dry Fucus, and made a fire of it upon the sands, and among the ashes coarse fragments like glass were found. This suggested an idea, and, by degrees, sea-weed ashes, or *kelp*, became a very important article of commerce, as an ingredient in the manufacture of glass, bringing in a revenue of £200,000 per annum to Scotland. The duty has since been taken off foreign barilla, which is better adapted than kelp for glass-making, &c.; and now the principal use of the sea-weed ashes is for the production of iodine. The properties of iodine were first ascertained by Gay, Lussac, and Davy, about 1815. Iodine exists in sea-water, and in sea-mollusks and weeds. In the chemical process for obtaining it, it is discharged from the kelp in a violet vapour, which is received into glass baloons placed for the purpose, where it becomes condensed into a solid black crystalline body. It is a most valuable medicine in scrofula and all swellings, and is of great use in photography. Martin Tupper alludes to the medicinal property when he says:—

"The sea-wort floating on the waves, or rolled up high along the shore,
Ye counted vile and useless, heaping on it names of contempt;
Yet hath it gloriously triumphed, and man been humbled in his ignorance;
For health is in the freshness of its savour, and it cumbereth the beach with wealth,
Comforting the tossings of pain with its violet-tinged essence,
And by its humble ashes enriching many proud."

The waste left by the iodine is excellent for manure, and the weeds are gathered in quantities to strew upon the land in their simple state. Their value is so well understood in the Isle of Arran that the factor of the Duke of Hamilton allots a portion of the shore to each person farming the adjacent land, and they carry the weed inland for manure; and also, in seasons of scarcity, for food for the sheep and cattle. The F. vesiculosus is called Swine-tang in Gothland, and pigs are fed principally upon it. Inglis, in his work on the "Channel Islands," gives a most interesting picture of the sea-weed harvest in Jersey. On the 10th of March and on the 20th of July, all the inhabitants repair to the shore, with such vehicles as they can press into their service, and they gather the weed in great quantities. Some they dry and use for fuel, and the remainder is laid upon the land as manure. It suits potatoe land remarkably well, and on this account is eagerly seized upon in Ireland, where it is carried at least fifteen miles inland. The Icelanders cook several of these olive weeds as food; and in Borneo there are some species entirely soluble, from which a strong jelly is formed and exported to China. The Fuci grow with wonderful rapidity: the Carr rock in the Frith of Forth was clear in November, and in the May following there was a dense crop of Fuci upon it, some of the plants six feet in length!

The Fucus order contains one more family, the sole British representative of which, the Sea thong, was not present in the heaps we were examining. Its peculiarity is that it grows first in the form of a top, and from that, long strap-shaped fork fronds shoot out, which merely live to ripen the seed, and then fall off. Edward got a fine specimen last February on the Coast of Cornwall, (Himanthalia Lorea, *Plate I., fig. 9*).

Leaving the dry weed on the shore we repaired to the margin of the water, and disregarding the danger of wet feet, we took our station on some low rocks covered with Fuci, at the foot of which noble fronds of the Sweet Tangle were idly waving in the calm tide. Here were specimens of the second order of Algæ the Laminariæ olive weeds bearing the spores in patches upon the fronds.

A frond with a mid-rib which we drew from the water proved to be the Edible Alaria (A. esculenta, *Plate I.*, *fig.* 5), it grew upon a stem, the main frond being a continuation of the short stem, while smaller fronds grew like wings from either side of the stem, it was formerly much used as an article of food. Gigantic tongue shaped fronds six feet in length, with curled edges and the middle puckered, were floating in every direction. These were Sweet Tangles (Laminaria Saccharina, *Plate II.*, *fig.* 1). sometimes called Sugar Sea-belt. This and many other species contains *Maunite*, a sweet nutritious substance resembling sugar, it is constantly eaten in Iceland, and in Japan it is reckoned a great delicacy.

The finger Laminaria (L. digitata, *Plate I.*, *fig.* 6) was also there, its fronds are divided into segments like gigantic fingers, the substance is so tough that we tried in vain to tear it, and after pulling with our united strength at one enormous frond for some time it suddenly came from its moorings, but brought with it a large stone which the fibres of its root had firmly clasped. The stem was ornamented with delicate white zoophytes, and glazed patches of red brown, which proved to be a minute sea-weed. The white net-work spreading over both stem and frond was but the forsaken home of a zoophyte, but living zoophytes of iridescent hues, are often found on such weeds, as described by Crabbe.

"While thus with pleasing wonder you inspect
Treasures the vulgar in their scorn reject,

See as they float along, the entangled weeds
Slowly approach, upborne on bladdery heads,
Wait till they land, and you shall then behold
Myriads of living points ; the unaided eye
Can but the fire, and not the form, descry."

There is a species of Laminaria which grows in the deep sea, and has a bulbous root, and another, an inhabitant of sandy shores in the South of England, which only attains the height of a few inches. All the members of this family produce good kelp.

Some delicate feathery weeds were growing upon the Laminaria fronds which we gathered and reserved for quiet examination at our lodgings, and turning from the rocks we began to retrace our steps.

Tangled masses of Sea Whipcord (Chorda filum, *Plate I., fig.* 4), were brought up by the returning tide, some of these were hairy, but most were rubbed smooth by the action of the waves. This weed grows to the length of 30 or 40 feet, the frond is hollow, but interrupted every few inches so as to form chambers ; the air which fills these hollows, buoys up the plant as the air vessels do the Fuci : sailors term it "dead men's ropes," it is very dangerous in catching and detaining any floating object, and would be fatal to any swimmer who ventured amongst it, for the cords though thin are very tough.

A small order of more delicate weeds succeeds that of the Laminaria, they are remarkable for bearing little tufts of fine olive filaments on the frond ; they become soft on exposure to the air, and they possess the property of decomposing other Algæ with which they come in contact.

Edward found a member of this order in Cornwall, which is called after Desmarest the French botanist; it is a pretty slender plant with tufts of green hairs along its tapering fronds (Desmarestia Aculeata). There are two other species of the same family, but we found neither of them. In the same order there is a Sporochnus and a Carpomitra, but both are rare even in the south of England.

Plate 10.

CHAPTER XVI.

SEAWEEDS.

" Mark well yon sea-weed rooted on the rock,
The maddened surge assails its fragile form,
And yet it moves not, clinging with small hands.
An emblem flower, methinks, of steadfast ones,
Who dwell in peace amid earth's wild turmoils."

IN our second ramble along the shore at Oban, we came to some charming rock pools, which the retreating tide had just left open to observation. One of these was full of a thread-shaped olive weed; the fronds were above a foot long, branched, and beset with tiny shoots. It was rooted to the rock by a little flat substance. The continual crossings of the branches procure for the family the name of Netted Tube (Dictyosiphon), and this our one British species is called the Fennel-like (Fæniculaccus, *Plate II.*, *fig.* 2).

Thrown up upon flat rocks, we found a weed of the flat forked form of the Fuci, but slender and thin, and of a golden olive colour. Under a magnifier, its fronds have a netted appearance on the surface, hence it is called Dictyota; its specific name, Dichotoma (*Plate II.*, *fig.* 3), is on account of the forked fronds. Both these weeds belong to an extensive order called after them Dictyotaceæ.

The Sea Endive (Haliseris) was once given to me from the coast of Devon; it resembles the Dictyota, but has a distinctly marked mid-rib.

The Cutleria, a southern species named after Mrs. Cutler, varies from these by the tips of the branches being repeatedly torn.

The most attractive member of the order is the Peacock weed (Padina pavonea, *Plate II., fig.* 4); it is an annual peculiar to Jersey and the south of England. When my cousin was staying at Lyme Regis, there were some rock pools which she called her garden, and here the Peacock weed grew in abundance. The fronds are fan-shaped, from one and a half to four inches high, covered with down and zoned with shades of brown and green.

The Taonia, also named in allusion to a peacock, has a frond torn into fingers, which grow broader towards the end, so that the whole frond has a somewhat fan-shaped form. It is marked with waved lines, which have a relationship to the zones of the last-named species (T. atomaria). The fruit is in clusters upon the surface of the frond.

The Punctarias have tongue-shaped fronds, and the fruit is scattered all over them like dots, hence the name. The P. latifolia is olive green, and grows between tide marks on the south coast. The P. plantaginea is more generally diffused; it is dark brown, and grows on stones or on larger algæ. The P. tennissima is delicate and pale, very transparent, and living as a parasite on zostera. I have had specimens sent to me from Jersey.

The next family, that of the Asperococcus, differs in the

fronds being tubular instead of flat. The clusters of spore cases are mixed with filaments, hence the name, which means *rough seed*. The Compressed Asperococcus is occasionally thrown upon the coast of Devon. Turner's Asperococcus is found on stones on muddy shores, and the Prickly species is also native to the Devonshire rocks.

The order of Chordariaceæ contains but a few humble weeds, some parasitical, some mere incrustations. We found one of the most important members of the order, the Chordaria Flagelliformis (*Plate II., fig.* 5), growing on stones on the shore at Oban; it resembles the Dictyosiphon Fœniculaccus, when its fibres are worn off. The Chordaria Divaricata has very irregularly divided branches, sometimes two feet long; it lives in deep water.

The Mesogloias have also much divided thread-shaped fronds; M. vermicularis is pretty frequent on our shores in summer, the branches are thicker in proportion to their length than in the allied weeds, and are slimy to the touch.

The Leathesia tuberiformis we did not find till afterwards, like an empty grape skin; it was lying among rocks at Arran; its globular form makes a strong contrast with its allies. It is named after Mr. Leathes, a distinguished naturalist.

The Ralfsia is a mere incrustation, a marine lichen, named after Ralfs, the botanist, dark brown in colour, and tough in texture.

A quantity of coarse olive weed, which we drew from the sea at low water, furnished us with specimens of the minute Elachisteas. The fronds in this family consist of a tuft of simple threads growing from a little knob. The Fucus Vescicularus was covered for inches with such tufts, they are called from the plant that nourishes them, Elachistea Fucicola, the generic name meaning *least*. Other species are found on the Cystoseira Fibrosa and Hemanthalia Sorea, etc.

The Myrionemas, the last family of the order, are formed of numerous erect threads, so small as to need a microscope to distinguish them. The whole plant forms a mere dot, and is found on red and green weeds.

The last order of the olive weeds, the Ectocarpaceæ, contains a number of elegant plants, some simply branched, and some most copiously divided, and coloured in every shade of olive and green: but it was not at Oban that we made the acquaintance of this family.

We had passed on to Arran, and achieved what we had been assured was an impossibility, viz., the procuring of lodgings at Brodick Bay. The Duke of Hamilton, whose castle stands on the left of the said bay, discourages summer visitors, and will not allow his tenants to enlarge their houses for their reception. But we cared for no luxuries, and were well contented with mere cottage accommodation, cleanliness being our only *sine qua non*.

How exquisite was the moonlight view over the frowning mountains, peaceful bay, and silver sea beyond! It was just the scene to suggest the "Seaweed dream."

> "I walked one night in dreams on the shore,
> And heard the mighty breakers roar;
> Yet amid the din there was borne along
> A syren sound, the sea-weed's song.

Oh, children of ocean, blythe are we,
Born in the depths of the brimy sea ;
Nursed by the motion of sounding waves,
Deep in the shadows of coral caves :
   Down, down
  In the dark blue sea.

Blythe children of ocean sure are we,
When first from the rocks our roots are free ;
And spreading our leaves we float away,
Up to the light of the living day :
   Up, up
  In the sun's bright ray.

Lone children of ocean, lost are we
When cast on the shore by the angry sea,
If none will gather the sea-weed spray,
To mix with the bloom of the garden gay ;
   Still, still
  To breathe of the sea.

So I gathered in haste the seaweed spray,
And brought it forth to the light of day ;
And I thought, as I did so, of other flowers,
Gathered from earth for heavenly bowers.
I pictured the beauteous spirit forms,
Once nursed below amid life's rude storms ;
And I thought of their blooming on Canaan's shore,
To be tossed by angry waves no more."

That night the voice of prayer and praise arose from the cottager ánd his family, mingling with the distant roar of the ocean, and the ceaseless soughing of the wind ; and we felt as if our little chamber were a mere bench in the great temple, where the mountains worshipped, and the trees bowed down their heads, and the sea uttered its solemn response, appealing most touchingly to "all that have breath," to "Praise the Lord."

The first forenoon after our arrival found us on the rocks below the castle, diligently seeking weeds. At first we but picked up an occasional new one out of the clear pools, but after a time we came upon a little plain of sand among the red rocks, where ebbing waves had deposited a perfect carpet of sea waste. These were surely the weeds that had called themselves "lost," in their song, and so the greater part of them remained: for it was but a small portion that we could rescue from oblivion. Yet what is "lost?" It is a good saying, "That is not lost which a friend gets;" and so the weeds which went to enrich the neighbouring corn fields were but little to be pitied.

First we seized upon a stiff shrubby weed, which gave us great trouble in drying; "floating out" it wholly disregarded, and we only overcame it at last by putting it between strong paper, with a large stone on the top of it. The branches were dark olive brown and woody, and adorned at each line with a little tuft or whorl of forked threads. This was the Whorled Cladostephus, (*Plate II., fig.* 8.)

Near it lay a similar weed, with more clumsy but softer branches, the thread-like foliage being longer and crowded all over the branches. On closely examining it the threads seemed quite simple; its appearance agreed with Harvey's description of the Spongy Cladostephus. Thus we were in possession of both the British members of the first family of Ectocarpaceæ.

Another dark weed lay among that mass of wreck, its

numerous, to the unassisted eye, simple branches, were straight, and arose from the same point in the main stem, thus forming a large tuft an inch long, and nearly as broad; and lower down on the stem other such tufts were planted; this peculiarity of form enabled us to identify our little weed as the Broom-like Sphacelaria (*Plate II., fig.* 9), a member of the family succeeding that of Cladostephus, and distinguished from it by the little branches or threads being pinnate, as the lens revealed those on our weed to be in a double degree.

Pretty tufts of dark brown were lying there with feathered branches about an inch in length which made beautiful pictures when floated out, these were plants of the Feathery Splacelaria (*Plate II., fig.* 10).

Edward had found the Desmarestia Aculeata in Cornwall, the preceding February, beset with delicate olive tufts but a few lines long, the minute branches simple in form, but revealed to be pinnate, that is with a row of threads on each side, when seen through a magnifying lense. These were plants of the hairy Sphacelaria.

S. Sertularia, Fusca, Radicans, and Racemosa did not come in our way.

One more weed rewarded our search that day. It was olive brown, and much entangled. The slender branches were so interwoven that we could not disentangle them even when they were floated in lukewarm water. This was surely the Woolly Ectocarpus (E. tomentosus), and it had a decided similarity to the delicate slender branches which we had gathered off the tangles at Oban. These we now floated out, and found that we had two species, both very elegant in form; the one a little more robust, more entangled, and less green than the other. The long slender branches looked like a filmy cloud in the water, but made a beautiful object on paper. The greener one we fancied was E. siliculosus, but as the lens revealed none of the pod-like fruit, we could not be sure whether it was that species or the E. litoralis (*Plate II., fig.* 11.) Fortunately Dr. Landsborough gives an ordeal by which these closely allied species may be distinguished, not that of fire, nor of red hot ploughshares, but of boiling water, a plunge into which turns the E. litoralis bright green. Thus we decided that the coarser plant was the E. litoralis, and the frail elegant olive green branches were E. siliculosus. There are many other species in this family, some with more simple branches, paler and of smaller size, but we were content with our representatives for a beginning. Of the Myriotrichias, with their densely crowded branched hairs, we did not meet with a single specimen, they both are mere parasites, living on Chorda, and are but a few lines in size.

Plate II.

CHAPTER XVII.

SEAWEEDS.

' The floor is of sand like the mountain drift,
And the pearl shells spangle the flinty snow ;
From coral rocks the sea-plants lift
Their boughs where the tides and billows flow.

The water is calm and still below,
For the winds and the waves are absent there ;
And the sands are bright as the stars that glow
In the motionless fields of the upper air.

There with its waving blade of green,
The sea-flag streams thro' the silent water,
And the crimson leaf of the dulse is seen
To blush like a banner bathed in slaughter."

PERCIVAL.

BEAUTIFUL as are many of the delicately formed olive sea-weeds, they become tame in our eyes when contrasted with the succeeding class, that of the red weeds. Although red is their prevailing colour, yet they vary to black, purplish, or brown, and their form and texture present endless variety. Now we have branches of the most complex and delicate workmanship, and now simple expanded fronds, or perfect crimson leaves. Sometimes the texture is so delicate as to tear with the touch of a camel-hair pencil ; sometimes it is

K

tough and leathery, and sometimes quite stony, with a coat of lime.

In search of any members of this group which the waves might wash to our feet, we set off early one morning to walk from Brodick to Corrie, a fishing hamlet on the shore, some three miles to the north. We passed quickly by the red rocks in the Castle Bay, because we had previously examined them; and we did not begin our search till we reached some similar rocks of red sandstone, midway to Corrie, where a vast congregation of gulls seemed to be taking their morning meal; and, at the same time, discussing the affairs of the nation. They took to the water at our approach, still uttering their sharp cries as they rode indolently upon the white crested waves. They had cause for annoyance, for the rocks from whence we had scared them were covered with periwinkles, mussels, and other small shell fish; and the heaps of empty and broken shells showed that the gulls were adepts at opening their own bivalves : they had no need of an oyster knife.

Amongst these rocks, also, we found a great quantity of sea waste, covering the miniature plains as with a thick carpet, and we could have filled our vasculum with Sphacelaria, and Cladostephus. We did stow away some pretty green weed for future examination; but we rejected the olive, and kept our attention for the red.

Leafy fronds, not very unlike a Fucus, only bluntly toothed at the margin, and of a dark red instead of olive green, were there; and we recognized in them specimens of that essentially northern weed Odenthalia dentata

(*Plate II., Fig.* 1), both the generic and specific name of which signifies *toothed*.

There, too, were plants of the Carrageen moss, growing in the low water pools, and crowned with long tufts of an elegantly branched red weed, shading to brown. The thread-like form of the branches, dark in the approach of winter, and which became darker in drying, proved it to be the Rhodomela subfusca (*Plate II., fig.* 2). The generic name signifies *red-black*, and is adopted as the family name of the first great order of red seaweeds, or Rhodospermeæ, not because of the importance of this Rhodomela and its brother, but because red-black is a characteristic of plants in the whole order. The R. lycopodioides has long branches, with clusters of fine branchlets here and there, which are much divided. We found it both here and at Ardrossan.

The Bostrichea Scorpioides, the one British member of the next family, is a dark weed, with prostrate divided branches, clothed with alternate threads, each of which curls round. Its name signifies a curl of hair.

The Rytiphlœas are so named because of their wrinkled surface; they are branched slender weeds, pinnate, and of a brown red colour. We did not find any specimens.

In the larger tide pools, and in wet places on the beach, attached merely to rolled stones, a bright crimson weed waved its long thread shaped branches, or reposed them on the wet pebbles, looking like a ruby stain. Some of the stems were as thick as a cord, and nearly a foot long; it was difficult to get them to stick to the paper in drying. This was one of the handsomest and most frequent of the large family of the Polysiphonia, a name derived from two Greek words signifying *many tubes;* the branches of the genus being tubular (P. elongata, *Plate II., fig.* 3).

An allied species of smaller size, harder texture, and darker hue, lying among the rubbish, proved to be P. nigrescens; and we found the wiry branches of P. fastigiata growing in abundant tufts on the stems and air vessels of Fucus nodosus. This last species was of a blackish brown colour.

There are numerous species of Polysiphonia, but these were all we met with on this occasion.

One little morsel of the beautiful Dasya coccinea (*Plate II., fig.* 4) crowned our search. Its name means *hairy* and *red*, and the main stem thread like, and clothed with hair-like branches fringed with the slenderest points justify the name. There are three other Dasyas, one the D. venusta, very rare; the others frequently found. This family is the last in the order of Rhodomelaceæ.

In the next order, that of the Laurenciaceæ, the colour is bright red, occasionally varying to pink or purple. The first family is named after the French naturalist Bournemaison, and contains only one British species, a beautiful plant, with regularly arranged, tapering, asparagus like branches.

The true Laurencias have thickish fronds, some round, some flattened; the L. pinnatifida we found under larger sea-weeds growing as minute brushwood on the sides of the rock; they were of a dark purplish crimson, pinnate and toothed; we did not find the Tufted or the Blunt Laurencia.

Neither had we the good fortune to find specimens of either of the Clavellosas, pretty weeds of a delicate pink hue and watery texture. Their name is in allusion to the golden hue the plants assume when steeped in fresh water.

Of the weed which we used to call "Crab's claws" we found plenty, both washed on to the shore, and growing in tide pools (Chylocladia articulata, *Plate II., fig.* 5). It has a thick round branched frond, drawn in to a very small size every few lines, so as to represent joints or articulations. It was very difficult to press for it would not lie as we wished it, and if we pressed it too much we crushed it. Edward found very large specimens of this on the Cornish coast last February, and also of the oval species (C. ovalis), where oval leaves are attached to a thickened stem. He got the C. kaliformis also; it is gelatinous and fleshy like the articulated species, but the stem is not tied in as if jointed. There are two other species which we have neither of us found.

Leaving these rocks we pursued our way towards Corrie by the road which lay between the shore and a strip of marshy ground bordered inland by another line of rocks. Upon this marsh belated flowers of the Pale Butterwort were still blooming amid masses of the Slender Bog moss.

The rocks forming the shore at Corrie are higher than those at Brodick, and the rock pools are larger and deeper. Many of these were carpeted with the common Coralline which was violet coloured in the deepest pools, and varying from pale lilac to cream white in the shallow ones. The stony coating of these Corallines induced naturalists to place them among zoophytes for a long time, and it has been only recently discovered that when the lime is cleared away by acid a vegetable structure is found beneath.

The delicate thread-like fronds of the red Jania (*Plate II., fig* 6), parasite on a withered morsel of Rhodomela were lying high and dry among waste ; slender as these much forked branches are, they are coated with lime, and of firm texture, (Jania rubens).

We were not astonished to find Nullipores on that coast, but it is always difficult to me to realize that the clumsy coral-like branches of the Melobesia polymorpha belong to the vegetable kingdom. Such is, however, the case, and we were glad to place the Nullipore in question among our collection of Algæ. There are other species of Melobesia, some growing on rocks in deep water, others incrusting shells, etc., and some so small as to form mere lichen-like patches on sea-weeds. The importance of this group is considerable, we are told by Dr. Landsborough that " the mortar used in the building of the Cathedral of Iona was formed of calcined shells, and a great quantity of the fragments of the white coral, which abounds upon the shore of the island," he further adds that it is easier to break the stones than the cement. He also gives a very interesting account of the bivalve Lima, who, having the same taste as St. Columba, builds himself a house binding together morsels of Nullipore with a cordage of his own manufacturing, and here he lives in safety among neighbours from whose voracity his slender shell would afford but scant protection.

The Hildenbrentia is a crustaceous weed like a dull

red stain on rocks at low tide. Edward met with it on the rocks at Looe at Cornwall but could not preserve it.

The next natural order is a very interesting one, containing the leafy Delesserias and Nitophyllums, it is called the Delesseria order ; both the family and the order are named in honour of the French naturalist, Benjamin Delessert. The D. sanguinea we found lying on the beach between Corrie and Brodick, the delicate crimson leaves torn as usual by the contact with the rocks. This weed is a favourite with every one, and as common as it is beautiful. I have seldom seen a collection, however small, which did not boast one of its rosy fronds. The D. sinnosa (*Plate II., fig.* 8) is scarcely less common or less beautiful, its fronds are varied as in the last species, and sinnated as in the oak-leaf ; it is often called the Oak-leaf Delesseria. We found small fronds and clusters of fronds of this plant both at Arran and Ardrossan, and Edward brought gigantic specimens from Cornwall in February. The Winged Delesseria (D. alata) is the most common red weed on the western shores of Scotland, its fronds are narrow and branched, and have narrow leafy edges or wings, the substance is gelatinous. We picked up an armful of it between Corrie and Brodick, and we could have done the same afterwards at Ardrossan. The species Hypoglossum and Ruscifolia are smaller and rarer species—the former with long narrow leaves, the latter with clustered broad ones.

The Nitophyllums are equally handsome Algæ, each frond consisting of a broad out-spread membrane of a full red colour ; the seed is dotted over the frond. We found specimens of the Torn Nitophyllum, the fronds of which are divided into forked lobes, and often edged with a fringe of tiny notches. N. punctatum grows to a large size in Ireland, being sometimes three feet broad. Dr Landsborough proposes it as a mantle for mermaids.

Only the pretty and familiar Plocamium coccineum (*Plate II.*, *fig.* 8), was now wanting to complete our Delesseria order, and the coast of Arran afforded us a nice specimen, and others were added afterwards at Ardrossan. This weed is the only one of the order with a thread-shaped frond, and it is plentifully branched, and each branch is set with abundance of alternate threads. The branches and threads are flattened, which makes this plant very easy to spread; and its beautiful rosy hue and elegant form, make it very desirable for collections and sea-weed pictures. Its name signifies *intertwined hair*, suitable because of the interlacing branches. Formerly this weed used to be so eagerly sought for as article of sale for the above mentioned pictures, that many poor people on the coast gained their living by collecting it. Certainly this family might appropriately use the lines of the poet—

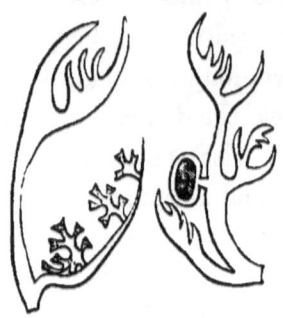

TWO KINDS OF FRUIT OF PLOCAMIUM.

> " Oh call us not weeds, we are flowers of the sea,
> For lovely, and gay, and bright-tinted are we.
> Our blush is as bright as the rose of thy bowers,
> So call us not weeds—we are ocean's gay flowers.
> And gay are our homes, 'neath the deep rolling waves;'
> Where we bloom mid the rocks and the coral formed caves.

There, unaided by culture, such bright forms have birth,
As vainly you'd seek mid the gardens of earth.
Not nursed, like the plants of a summer parterre,
Where the winds are but sighs of the evening air ;
Our exquisite, fragile, and delicate forms
Are reared by the ocean, and rocked by the storms."

CHAPTER XVIII.

SEAWEEDS.

" See on the violet sands beneath,
   How the glorious shells do glide!
O sea! old sea! who yet knows half
   Of thy wonders and thy pride!

" Look how the sea plants trembling float,
   All like a mermaid's locks,
Waving in thread of ruby red
   Over those nether rocks!

" Heaving and sinking, soft and fair,
   Here hyacinth, there green,
With many a stem of golden growth
   And starry flowers between."

WHEN we began our Highland tour, we agreed not to mind the rain; a wetting in the Highlands does no one any harm. We had found great peace of mind in this assurance. Thus when we set out at Arran for a coast ramble, and found that the clouds which had hung all the former day over the hills, had determined to empty themselves to-day, we still were resolved not to mind. But surely there is no rain like Arran rain! A steady breeze was blowing, and the rain came with it—not gradually, but all at once, a tremendous pour! We sheltered behind rocks, but the

Plate 12.

wind took the trouble to sweep round them, and bring rain enough to drench us in a quarter of an hour. We were thoroughly beaten, and obliged to drag our dripping persons home, where we were met by our kind helpful hostess with much friendly sympathy. In answer to my explanation of " we staid awhile hoping the rain would cease," she replied with the usual Arran rejoinder, " Just that, Mem," and then assured us that we must make up our mind to stay within doors, for that when once it began to rain in Arran, it seldomed " faired " that day. So we were compelled to resign ourselves to our fate, and after exchanging the wet suit on our backs for the dry suit in our carpet bags, we applied ourselves to spreading and arranging the weeds in our possession. Alas, the morrow was little more promising, but in the afternoon we made another attempt to get to the rocks. The rain had almost ceased, and the high wind was fast drying the roads. This time we turned southwards, taking the coast towards Lamlash. We each devoted ourselves to favourite pools, and were far separated, when a sudden gust whisked off my hat, at the same moment tearing away the comb round which my hair was coiled, and blinding me with my own mermaid locks; before I could even call to my companions, my confusion was augmented by another freak of the wind, which lifted up my cloak, and wrapped it round my head, blowing hard enough the while to threaten me with an entire overturn. As a caution against this impending danger, I sat down, then unfolded the head, imprisoning my hair under the hood of my cloak; after a long search I found that my comb had returned to its native element, and was nestling among branching weeds in a tide-pool; I reclaimed it, and once more conquered by Arran weather, I signalled to Edward to come to help me home—this was no easy matter, for the wind was becoming unmanageable, and my companions were so convulsed with laughter at the plight I had been in, that they were no assistance to me at all. This was our last excursion in that lovely island, and its reward was but a green weed or two, of which more anon.

The sands between Ardrossan and Saltcoats, and even the rocks, fertile as they were, looked very dull and prosaic after the ever-varying beauty of Arran, but our time was up; a day at Ardrossan, and a few days in Edinburgh, and our tour must end.

Upon stretches of sand among these rocks we found heaps of sea-weed lying, blown thither by the breeze of yesterday, which might have blown me thither also. We eagerly seized some large red leafy fronds of strong leathery texture, and very variable form. It was generally growing in tufts on uprooted stems of Laminaria Digitata, and was sometimes club-shaped, and sometimes split into forks or fingers. This is what the Scotch call Dulse, (*Plate XII., fig.* 1), and as such was formerly sold in the markets, and eaten either raw or cooked. Cattle are very fond of it, and seek it out eagerly when they can get to the shore; indeed I have read that sheep have often been lost by going too far out to seek it, and it was hence called Sheep's Dulse. Dulse in Gaelic means *leaf of the water*—it was formerly dried and chewed as tobacco (Rhodymenia palmata). This plant belongs to the family which gives the name to the eleventh order of

British Seaweeds, the name signifies, *red membrane.* The R. lacimata was also lying among the stranded weeds, the frond is thick and leathery, and cleft into many broad blunt divisions, the margin is fringed with tubercles. We also found the R. jubata that same day, its frond is egg-shaped and tapering, often clustered, and fringed with young fronds the same shape as itself. There are three other species, but they have not yet found their way to our infant collection.

The Sphærococcus is too scarce a weed in Scotland for us to have any hope of finding it. It is thick and gelatinous, doubly branched, and dark in tint.

The Gracelarias are also gelatinous weeds: they are slender and pretty, but pertain chiefly to the south coast of England. Dr. Greville states that a species of Gracelaria is greatly valued as an article of food in Ceylon, and that our G. compressa is little inferior to it. G. tenax makes excellent glue, and is used for this purpose by the Chinese, who employ the glue in glazing silk, gauze, and paper. Mrs. Griffiths prepared a beautiful pickle from the G. erecta, small specimens of which Edward found when in Cornwall.

The Hypnea purpurascens we sought in vain: it has since been given to me by a friend, who found it early in the season on the same coast. It is a slender weed, much branched, the branches either simple or forked, or thread-shaped.

The great order of Cryptonemiaceæ contains many genera, generally of a thread-like form and leathery texture.

The rare Grateloupia, named after a French naturalist, I have never found.

The Gelidium cartilagineum was sent to me from Jersey. It is pale and much branched. Its branches are too strong and firm to become entangled: and the flat pinnate branches of the very similar gelidium corneum were here upon the Ardrossan shore.

A species of Gelidium is said to be the plant chosen by the swallows of Japan for building their famous "edible nests," which form an extensive article of commerce with China. Burnett tells us that in his time, £230 to £290 worth of these nests were commonly exported from the Indian Archipelago, and were sold in China at the rate of £5, 18s per ℔. The collecting of these plants, according to Mr. Crawford, is as perilous as our samphire gathering used to be: they are found in damp caves, and are more esteemed if taken before the birds have laid their eggs. They are collected twice in the year. Some of the caverns can only be approached by a perpendicular descent of many hundred feet by ladders of bamboo and rattan, over a sea rolling violently. When the mouth of the cavern is attained, the perilous office of seeking the nests must often be performed by torchlight; and as you penetrate far into the caverns, the slightest trip would be instantly fatal.

In the rock-pools at Ardrossan, there was abundance of a red leathery forked weed, with little pimples all over it. It was very tough, so that it was not easy to detach it from the rock, and still less easy to press it for the herbarium. This was the Gigartina mammillosa (*Plate*

XII., fig. 2), common to all our coasts. The other species of the genus are rare.

In the same pools, the Chondrus crispus or Carrageen moss (*Plate XII., fig.* 3) was growing in great luxuriance, its tough forked fronds of a fine crimson hue bearing iridescent tints under the water, so that I plunged my arm in again and again, hoping to bring out a frond tipped with blue, but the lovely hue disappeared as soon as the weed left the water. Its name of Carrageen moss arises from the fact that its edible properties were first demonstrated at Carrageen, in Ireland. It has long been much esteemed as an article of food in the sister isle ; and when bleached in the sun, stewed down to a jelly, and mixed with wine or cream, and any flavouring, it makes a most nutritious and palatable dish. It is often sold by chemists as a substitute for Iceland moss, and is well worthy the attention of invalids.

Specimens of the Phyllophora rubens had been given to me by a Scarborough collector, but we found none on the Scotch coast. It is a pretty weed, with a variously-shaped leafy frond, which throws out young fronds from the margin. There are two other species of Phyllophora. The P. membranifolia is not uncommon.

The Gymnogongrus griffithsea is a native of Devon : its name means *naked wart*, in allusion to the form of its fructification. The Polyoides Rotundus is also remarkable for bearing large warts on its surface, in which the fruit is contained ; it was sent to me from Jersey. It has a narrow frond, repeatedly forked, and broadest at the tips.

The Furcellaria fastigiata (*Plate XII., fig.* 4) grows in

bunches; each frond as thick as whipcord; forked once and again, and tapering at the tips : there was plenty of it among the Fuci, on the Ardrossan rocks.

The Dumontia filiformis, with its thread shaped frond and simple branches, and the rose-coloured palmate Halymenia Legulata, we did not find.

The Kallymenia reniformis is a lovely weed; its name signifies *beautiful membrane*; the expanded frond is roundish and stretched at the margin; the specimen given to me was about three inches in height. The tide pools furnished us with the Tridœa Edulis, or Sweet Dulse; the frond is egg shaped, but tapering toward the base; it is dark red, and of a leathery texture. This species used to be much eaten both by human beings and cattle; and Harvey says it is still used as an article of food by the poor, either raw or fried.

In some of the shallow pools the little Catenalla opunita was growing under the shelter of large weeds; it looked like young plants of Chylocladia articulata, the branches being contracted at the base; the whole height of the fronds was but half an inch, and the colour was a brownish purple. Its name means *little chain*.

The Cruoria pellita, or Blood-stain, succeeds the Catenella in order; the glazed brownish stains upon the stems of the Laminaria digitata, which we had found at Oban, were plants of this weed.

No specimens of Naccaria, Gloioisphonia, Dudresnaia, or Crouania, came in our way; most of these weeds are rare, or confined to the south coast.

Of all the beautiful weeds on our shores, none can outstrip, and few equal in charm, those belonging to the last order of red algæ, the Ceramiaceæ. Composed of thread-like branches, each branch a string of cells of a texture and form most exquisitely delicate, the more we look into the structure of these plants, and the higher the power of the lens we use to aid our vision, the more we wonder and admire the amazing skill of their Creator; truly

> " Each rock pool has its treasure ; every tide
> Strews on the yellow sand from ocean's lap
> Weeds, than our flowers more fair."

We sauntered along the flat shore between Ardrossan and Saltcoats. As we approached a group of low rocks near the latter place, the waves washed a beautiful weed to our feet, whose slender twice compound branches, beset with minute thread-like foliage, bore a great resemblance to a plume of feathers. The leathery nature of the frond, its deep red colour, and thrice pinnate form, proved it to be the Ptilota plumosa *(Plate XII., fig. 7)*, one of the weeds more plentiful in the north than in the south. Quantities of Delasseria alalata, and sinnosa, were washed ashore by the same waves, as well as many other weeds.

> " See the glittering waves advancing,
> Crown'd with garlands from the sea ;
> Lovely water-spirits dancing,
> Wreathing ocean gifts for me."

The Ptilota sericea was there too ; it is a more slender species, and its texture is softer.

L

Another species that freighted those waves was the Ceramium Rubrum, a pretty bright red weed with tapering branches, the tips a little curled. It is a common weed, and very variable; our specimens were four or five inches in length. We found a lovely small Ceramium afterwards on rocks at Granton, in the Firth of Forth, at nearly low tide mark. The fronds were about an inch and a half high; the articulations very clear, giving the principal stem the appearance of a string of minute oval beads, ruby and white alternating; the tips were daintily curled in (C. acanthonotum); under a magnifying lens you could see three thorns planted outside each tiny branch. There are a great number of species of Ceramium, all very interesting; some as fine as hair, with hooked tips, and delicately marked articulations.

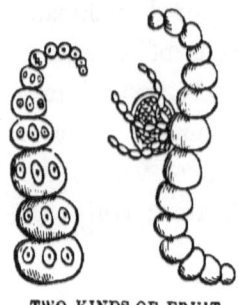

TWO KINDS OF FRUIT OF CERAMIUM.

The family of Griffithseas is equally beautiful; more robust in form, but very perishable in texture. The articulations are very plainly marked in this family also. They are named in honour of Mrs Griffiths of Torquay, who did much to facilitate the study of Algology; they are all inhabitants of the south coast of England and west of Ireland.

The last family of red weeds is that of the Callithamnion. They are delicately formed, and feathery; with stems either cellular or jointed; and sometimes transparent. The C. plumula we found in the Firth of Forth; the tiny stems were much branched, each branchlet resembling the end of an ostrich feather. There, too, was the C. arbuscula, the thicker opaque stem, naked in the lower part, and then bearing crowded branches, gives it a striking similarity to a miniature tree. The C. spongiosus retains water like a sponge; it grows parasitic on other algæ, on the north coast of Devon. The C. roseum prefers muddy shores; its fronds are fan shaped. C. turneri grows as a creeping parasite; and C. polyspermum is to be found on the coarser algæ. C. brodiæi is only found on the coast of Northumberland, and the south of England; whilst C. pedicellatum is dredged from deep water. None of these species favoured our collection. A wet day at Ardrossan was well spent in spreading our beautiful red weeds, and very creditable our collection looked; we had green weeds too, but we were not quite ready to study them. Our love for the seaweeds had so increased that we had not cared to cast a look inland at Ardrossan, we saw no beauty but in algæ.

> "Oh, I love the ocean flower,
> Gem of the unbounded deep;
> And through many a future hour
> Will the fond memorial keep.
> It tells that in the mystic world,
> Deep where resistless waters flow,
> Where the wrecked barque is wildly hurl'd,
> Untrodden fields and forests grow.
> As from the green and sunny land
> Gems of richest beauty spring;
> Form'd by the same unerring hand
> Comes forth the ocean offering."

"The tufted seaweed slenderly
Hangs from its place of birth;
And the blue waves left it tenderly,
To kiss the green lips of earth."

CHAPTER XIX.

SEAWEEDS.

THE third order of Algæ, the Chlorosperms, are distinguished by their green colour, though there are exceptions in favour of a more or less brilliant purple. The spores are formed within the cells, of which the whole plant is composed; indeed all the colouring matter is capable of becoming reproductive. These spores, when ripe, frequently move about, as if endowed with life, because of threads which are kept in motion by the water. We will turn our attention to the marine species first, and then make the acquaintance of a few of their fresh-water relations.

It was on the coast of Cornwall that I saw the first family in the Chlorosperm tribe. The Codiums belong to the Siphoneaceæ order, the characteristic of which is their tube-like branches.

I shall not easily forget the scene of grandeur which burst upon my vision as, after threading the narrow streets of East Looe, more like back alleys than important thoroughfares, and passing by the church, I came out on an open space of ground, girt with rocks, and commanding a full view of the wide ocean—

"'Tis the great Atlantic sea!
Many coloured floor of ocean, where the lights and shadows flee;

Merry billows running landward, with a sparkle and a song,
Crystal green with foam enwoven, bursting, brightly spilt along;
Thousand shapes, of living wonder, in the clear pools of the rock;
Lengths of strand, and sea-fowl armies, rising like a puff of smoke,
Drift and tangle on the limit, where the wandering water fails,
Level faintly-clear horizon, touched with clouds and phantom sails."

A spring tide was rolling in, ever and anon depositing heaps of wreck, which lads were gathering eagerly, and carrying away in donkey carts. I, too, claimed a share of ocean's bounty; and carrying a good bundle to a rocky platform, out of reach of the spray, but in the midst of the musical roar, I examined my treasures.

There were large sprays of the Oak Delesseria, all beset with colonies of young fronds, and masses of the Jointed Chyloeladia, looking like branching strings of glossy beads. The Rose Delesseria was there too, and abundance of the Dulse, and many another old friend, but the prize specimen was a frond of the Downy Codium, a foot in length! (C. tomentacum, *Cut I., fig.* 1). The branches were forked and cylindrical, about the thickness of a common cedar pencil, and covered with soft down. The colour was myrtle green, and the texture of the weed was spongy. I have seen specimens from many places on our shores, but never any so large as my Cornish treasure. The purse-like Codium is a great contrast to its tall brother. It is a perfect hollow ball, a little drawn out where attaching to a rock; its texture, too, is spongy (C. bursa, *Cut I., fig.* 2). It is very rare; my specimen came from Jersey. There are two other Codiums, but they are little more than incrustations, and are also very rare.

We sought in vain for the Bryopsis on the shores of Scotland. The feathery species is found at Portobello, but we were too late for it, it being a summer weed. Subsequently, however, I had occasion to spend a day at Hastings, and secured an hour or two for the rocks between that place and St. Leonard's. The sun was shining brightly, and the marine gardens in the rock pools wore their most attractive colouring. Heavy masses of the Rock Cladophora clothed the upper part of the rock basins; but beneath their sturdy foliage I noted transparent green branches, waving to and fro as some minute crab or fish disturbed the waters. Securing some of these, I found I had gained possession of the much desired Feathery Bryopsis (B. plumosa, *Cut 1., fig.* 3). The fronds were three or four inches in length, branches issuing from each side the stem at regular intervals, and beset, in their turn, by little branchlets. The weed was yellow green, and transparent, and looked quite glossy when dried. Professor Harvey states that if the point of one of the branchlets be wounded, the colouring matter of the whole frond may be pressed out at the aperture, thus proving, by a simple method, that the whole frond is formed of one tube. I found this experiment succeed entirely. There is a Hypnum-like Bryopsis, but we have not found it. It is more slender, more compound in its branches, and smaller, a more moss-like weed.

Vaucher's seaweeds come next (Vaucheria); their fronds are thread-shaped, and tubular, and they have spores attached outside the frond. There is a marine species with forked fronds, two or three inches high, found at Weymouth; and a submarine one with longer fronds, parasitic on red seaweed; and a velvet-like species, growing as a coating on muddy shores, but we have found no specimens of these.

The Conferva order comes next, taking its name from a Latin word meaning to consolidate, because some of the old practitioners made use of the true Confervæ to bind up broken limbs, a use to which their softness and power of retaining moisture well adapted them. The plants in this order are formed of simple or branched threads, green in colour, and articulated.

The fronds of the first family (Cladophora) are much branched. The Rock species (C. rupestris, *Cut I., fig.* 4) was growing abundantly in the tide pools at Hastings, and we had also found it in similar places at Arran and on the Firth of Forth. The colour varies from dark green to olive, or pale sage. The fronds are bushy, with numerous rigid branches, these again clothed with branchlets, pressed close to the stem. It is a pretty weed for ornamental work, forming, with zoophyte cases, shells and other weeds, pretty groups to place under a glass shade.

The Angular Cladophora (C. rectangularis, *Cut I., fig.* 5), is a very rare species, only thrown up during the summer months in southern shores. The specimen in our collection was found by Mrs. Griffiths. The more slender fronds of C. diffusa, we found at Arran, but slightly branched, and only tufted towards the tips. C. gracilis we also found there in rock pools, growing in dense tufts, the thread-like branches waving in the water, and looking like a film of yellow green. The colour faded somewhat in drying, but the specimens continued very silky. The C. albida grew near it, its long slender branches densely crowded. These were all the members of the large family of Cladophora which we succeeded in finding.

Arran had provided us with three Confervæ, the family characteristic of which is the simple thread shaped, unbranched fronds. They are all articulated, some very evidently so, and the fruit is contained in the articulations. The first Conferva we found was like a lock of very coarse horse hair; of a dull green colour, curled, and much entangled; it was on swampy ground, where a little rivulet discharged itself, and the tide occasionally washed over; it was Conferva litorea. We also found the long, straighter and less clustered threads of C. melagonium, (*Cut I., fig.* 7), and a green coating on the other weeds proved to be the delicate twisted threads of C. tortuosa, (*Cut I., fig.* 6). This family completes the order of Confervaceæ.

That of Ulvaceæ succeeds it. We have here broad expanded fronds as a rule, and thread-shaped ones as an exception.

The family of Enteromorpha have tubular fronds. E. intestinales is a handsome weed, with fronds sometimes a foot and a half long, and an inch broad; it is of a full green when fresh. E. compressa is very common, we found its fronds in tide pools at Oban, Arran, Ardrossan, and the Firth of Forth, varying from a thread to a half inch in breadth. It is distinguished from the former species by being branched, while the fronds of E. intestinalis are always simple. The fronds spring from one base, and gradually widen; they are blunt at the ends.

E. clathrata, forming an entangled silky tuft of narrowest threads, and E. ramulosa, as intricately tufted, but rough to the touch, were both adorning the rock pools at Arran. Of the other species we have no examples.

The Ulvas have flat outspread fronds. The common Ulva (Lattissima, *Cut I., fig.* 8), is bright, green, and much waved; it grew plentifully on rocks at Granton, making them dangerously slippery. The U. lactea is smaller and paler. The narrow, U. linza we did not find. The Porphyras were there also; the P. laciniata with its divided fronds, and the P. vulgaris with its large simple ones, both which are purple and glossy, and are called by children "sea silk." These plants make a wholesome dish when cooked, and are sold for that purpose by the name of Laver in various districts. According to Lightfoot, Ulva latissima is employed in the Scottish Highlands to bind about the temples in fevers, and is thought to induce sleep; and in the Western Isles it is stewed with pepper, vinegar, and onions as a dinner dish.

These Ulvas are most ornamental plants in tide pools and shallows: their bright green hue setting off the crimson of the Callithamnion, and the lilac the Coralline to the greatest advantage.

> "How varied the hues of marine vegetation,
> Thrown here the rough flints and sea-pebbles among;
> The feathered Plocamium of deepest carnation,
> The dark purple Sloke and the olive Sea-thong."

The Bangias are minute weeds; they vary, as the Porphyras do, from the established green of the class, to bright purple; they are little more than encrustations.

The last family of the Cladospermeæ, that of the Osyillatoriæ, is composed of very minute members. They are formed of small jointed threads, each composed of a simple tube. The name Oscillatoriæ is given to the order because of the constant movement of the fronds.

The first family in the order is called Rivularia, because many of its members inhabit rivers, of which more anon. We found one of the marine species, the Shining Rivularia (R. nitida, *Cut I., fig.* 11), on rocks at Arran, which it rendered terribly slippery. It grew upon the flat part of the rocks filling every tiny crevice, and plentifully sprinkling the fronds of the Rock Cladophora. The fronds are round, black green, and glossy, seldom as large as a good sized pea. All the species in this family are more or less globose, the threads composing them being woven together into that form.

When we came to examine some of the red weeds we had gathered in such abundance at Ardrossan, we found them beset with little tufts of green threads. The simple form of the thread-like fronds, and their veins devoid of motion, pointed out the little parasite as belonging to the Callothix family; its glaucous hue fixing the species as the Conferva-like (C. confervicolor, *Cut I., fig.* 12). Other members of this family grow upon rocks near high-water mark, or are parasitic on different algæ.

Lyngbye's sea-weeds are green, with thread shaped fronds marked with rings more or less clearly. They greatly resemble Confervæ, and their fronds are disposed in layers upon rocks, mud, sea-weed, or floating timber.

Thus in a comparatively short season we gathered together a fair number of seaweeds. We had Olive weeds

(Melanosperms) of every marked family; Red weeds (Rhodosperms) in abundance; and Green weeds (Chlorosperms) in plenty. We had seen them in their homes, gazing with glad admiration into many a fairy conservatory blooming in a tide pool, or peering into the forest-depths of ocean from a boat; or we had waited amid roar and spray for the crested waves to land their merchandise, and seized the dripping treasures as they were thrown upon the shore. Often did we ask with the poet—

> " Is there a quiet world that lies beneath
> The might of waters, sounding, moving ever,
> Or rough with storms, or thundering on the shore
> With deafening clamour? Yes, a quiet world
> Doth lie beneath, with groves, and vales, and streams,
> And living creatures, each with haunt and home,
> As best befit it."

But we must not overlook that seaweeds have a past history, a record graven in the rocks. In the oldest system of fossiliferous rocks, the Silurian, black marks of seaweeds are found; and more distinct traces of an Algæ allied to the Sea Whipcord, not a dangerous weed in those days, when all breathing creatures had their home in the waters. The Carrageen and the Fuci had also their representatives in those ancient seas; so that Algæ can trace back their family lineage further even than the Ferns and their allies.

"The earth is full of God's goodness," this we heartily assent to, because we see it with our eyes, and have daily experience thereof. But though less patent to our ken, yet just as surely, "So is this great and wide sea, wherein are things creeping innumerable, both small and great." The healing Iodine, the nourishing Carrageen, the masses of weed so important to the soil—all these are God's gifts from the bosom of ocean which we can understand and value—but it will require ages to reveal the full meaning of His love in His profuse furnishing of the world of waters; perhaps indeed we may only discover it fully when "there shall be no more sea."

CHAPTER XX.

FRESHWATER WEEDS.

> "And there, to charm the curious eye,
> A host of hidden treasures lie,
> A microscopic world that tells
> That not alone in trees and flowers
> The spirit bright of Beauty dwells,
> That not alone in lofty bowers
> The mystic hand of God is seen,
> But more triumphant still in things men count as mean."
>
> GARDENER.

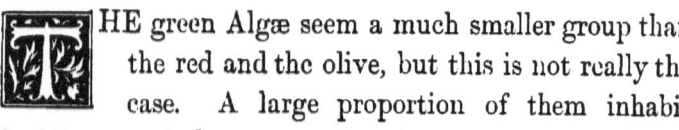HE green Algæ seem a much smaller group than the red and the olive, but this is not really the case. A large proportion of them inhabit freshwater and damp ground, and are therefore separated from the sea-weeds, though nearly allied to them in nature; and a vast number again belong to the entirely microscopic orders, Desmidiaceæ and Diatomaceæ.

In the same group with our marine friends, Codium and Bryopsis, we find the extensive family of the Vaucherias, so called from their historian Vaucher. The three salt-water species are very inconspicuous. The Forked Vaucheria (V. dichotoma) is a much larger plant. Its thread-like branches attain a foot in length; they are of a blackish green, and float in stagnant water. My specimens are from the neighbourhood of Edinburgh, and were gathered in the summer. Dillwyn's Vaucheria is very common. I first saw it at Hawkhurst, in Kent. Heavy rains had fallen continuously for several weeks, and the hop-growers were on the verge of despair —the clay soil held the water, and the lawn became a morass. After awhile I saw dark green threads growing in tufts at the foot of the grass-plants, and gradually forming little mats around them. Their tiny stems were irregularly branched. It was the first Vaucheria I had found, and greatly did I rejoice over it. The Budded Vaucheria (V. gemmata), so called from the buds growing on the sides of the branches, and present in a less degree in the other species, is to be seen frequently in dense floating masses. It is of a light green; its long branches are very much entangled. Ponds about Hawkhurst furnish abundant specimens.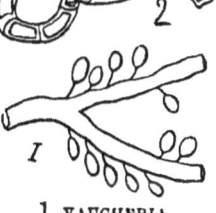

1 VAUCHERIA.
2 CONFERVA.

The Bird's-eye Vaucheria, (V. ornithocephala) is of a brownish colour, much branched, and forming elegant tufts. It grows in pools and ditches in autumn. My specimen came from Ecclesfield. Ditch banks in Kent and Wilts have furnished me with the ground Vaucheria (V. terrestris). It lies like green coating on the earth, and is rough to the touch, being beset with bristles.

The Grain-like Botrydium is a near relative of the

Vaucherias. It resembles little green seeds, these being vesicles with tiny roots. I found it in a turnip field at Hawkhurst that same wet summer. Some of the vesicles seemed to have opened, and discharged their contents; for they were cup-shaped. All these plants belong to the Siphoniaceæ order.

Two curious plants rank early in that of Confervaceæ, though at first sight they seem to bear little resemblance to any other in that thread-shaped colony. The two Chætophoras are gelatinous-looking plants, but the characteristic threads are present nevertheless, but embodied in jelly. The Endive-leafed species is found in streams. It grows to the height of two or three inches, with forked or indented fronds. The Elegant Chætophora grows in ponds or slow streams, attached to stones or stems. It is of a full green, round, and varies in size from that of a pea to a hazel-nut.

The Draparnaldias are pretty little plants, green, elegantly branched, and about an inch high. The Globular species has a beaded appearance from the contraction of the stems. The Dwarf one is feathery in its form. I have not found either of them, but have received specimens.

It was in a wood near Richmond that we first saw the orange Chroolepus: a moss-covered wall, under the shade of thick trees, was gaily painted with it, the patches resembling pieces of loose felt of a full orange colour. The naked eye can discern the threads of which the mass is composed, but we had to use a microscope to see the branches. Of course we pressed specimens for our collection, and were mortified to find that they changed their brilliant colour for a pale sage green. There is a violet-scented Chroopelus (C. jolithus), of a purplish-red colour, appearing as a coating on rocks and stones. A curious superstition attaches to this Alga in Wales. It grows on the stone in and about the well of St. Winifred. The legend states that the saint was flying from Caradoc, who, overtaking her as she neared the church where her parents were, drew his sword, and cut off her head. The head rolled into the church where St. Benno was preaching, and he, the saint, picking it up, fastened it on, and the maiden was none the worse. She survived this startling accident fifteen years, and became abbess of Gwythern, in Denbighshire ; but the cruel caradoc dropped down dead on the spot where he had committed the impious deep. A well opened where the head of St. Winifred had fallen, which is said to throw up fifteen tons of water every minute, but all these waters do not efface the marks of the tragic deed ; for

"In the bottom there lie certain stones that look white,
But streaked with pure red, as the morning with light,
Which they say is her blood !"

Numerous as are the marine members of the true Conferva family, they are equalled by the freshwater ones. On one occasion when I was rambling on the banks of the Wye, near Ross, after a long season of drought, I found something very like a seaweed adhering to stones in the shallows. By carefully picking my way across the partially dry river-bed, I reached a pool where this same weed was floating freely, its long green branches twisted like hair, and attaining a length of nearly half a yard. This was the River Conferva. It grew in company with the Common Chara, the Water Thyme, and the Greater Water moss. The dingy Conferva (C. sordida) is equally common. It is of a dirty green, and forms a cloud round the stems of pondweeds. It grows in great abundance in some fish-ponds at Elfords, near Hawkhurst. In the same neighbourhood I found the Entangled species floating in a yellowish green mass on half-dried ponds during the summer months (C. tracta). Its filaments are branched.

I have frequently found the Chestnut Conferva entwining among mosses, or forming a coating on a bank side. There is a black species, and a violet one with very fine branches, but these I have not even seen.

The Ulvaceæ order has also its fresh water members, one, which floats like green bubbles upon stagnant ditches, is familiar to every one, only it never struck most observers to call the noisome coating an Ulva.

The order succeeding that of Ulvaceæ is entirely composed of freshwater Algæ. Being very minute plants, they are of course described by an extra long name, *Batrachospermeæ*. The name means Frog Spawn, and all the members are slimy to the touch. The River Lemania is a sturdy looking little plant, of an olive colour, and with the branches circularly bent.

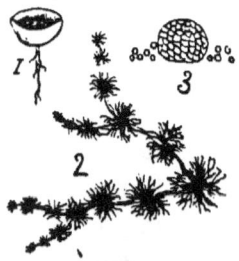

1. BOTRYDIUM.
2. BATRACHOSPERMUM.
3. PROTOCOCCUS.

There are many deep muddy ponds about Hawkhurst, but there is one very small one famed for its clearness. Searching for mosses one day, I repaired to this pond. Having noticed the fronds of the greater Water-moss waving in its limpid waters I secured my prize, little suspecting that a far rarer treasure was growing by its side. Peering once more into the depth before rising from my knees, I espied what I believed to be a fine filmy moss, and I plunged my arm in again to secure it.

I grasped cautiously and felt the stems snap, but the branches slipped through my fingers as if they had been live fish, and only one frond was coiled round my fingers when I drew my hand from the water; but that frond was of such excessive beauty, that I was overjoyed by its acquisition. It resembled a string, or rather a series of strings of the most delicate beads, branching in a pinnate form; the thick part of the stem was composed of beads as large as rape-seed and olive in colour, but the tips of the branches were delicately fine, and their beads purple. I soon perceived that many of the fronds that I had detached were floating about in the pond— some of these I caught on the end of my umbrella but I could not hold them, they were so slippery. After struggling for some time to circumvent the plant, I allowed it to circumvent me, and left it victor of the field, or rather of the pond; but returned with a sea-side tin on the morrow, and secured abundance of my pretty Alga. This was the common Batrachaspermum (B. vagum.) If I admired its beauty when seen with the naked eye, I did so much more when examining it with the microscope, where every bead appeared as a lovely round cluster of delicate branches. There is a black species and a green species, both delicately beautiful, and resembling this one in structure.

Flood-left pools by the side of the Wye furnished me with specimens of the Zygnema Quininum. Its fronds are thread-shaped, and entangled, and the masses were so slippery to the touch, that I thought they should have belonged to the family named after the frog-spawn. The Dotted species has the same slippery texture, indeed it is the characteristic of the family. The Bent Zygnema (Z. genuflexum) is finely branched, mud coloured, and only about an inch high ; it is scarcely possible to clear it from the mud among which it grows. A very refined lady was one day greatly amazed to see some of these mud-dwellers washed and floated out. She loved all created things, but a numerous family of young children left her no time to study minute herbs. Speaking of a well-known naturalist, she said, " when she takes up a bit of mud, shakes it in the water, and puts a bit of paper under it, a beautiful Alga appears ; but if I pick up a bit of mud, and do the same, there is only a bit of mud on the paper."

Upon stones, in streams in Swaledale, I have found dark green bodies like small peas ; these were plants of the Freshwater Rivularia, a genus belonging to the Oscillatoriaceæ group. The Echinellæ succeed the Rivulariæ, they are minute weeds growing in freshwater, and parasitic in Confervæ, and other water plants. The Circular species grows round the stem, radiating like the spokes of a wheel; the Bundle species is in clusters of club-shaped fronds; while those of the E. paradoxa are thread shaped, and branched, and thickened at the ends. None of these plants exceed the eighth part of an inch in height.

The Scytonemas are found both in the sea and in damp places. Matted together, gelatinous when wet, its members are found on rocks, hills, and banks. One of them resembles mouse-skin. The Bird's-eye species looks like velvet-pile on rocks. The Oscillatoriæ proper are formed of simple threads. The Dark Green species floats cloud-like on the surface of ponds, the tuft two or three inches across. The Pale green one frequents wells. The Ochraceus species is found in pools and bogs, looking like a brownish cloud traversed by very fine threads. The bark species is like a skin on damp wood, when dry, pealing like glaucous riband. There is a blue green species on damp ground about old buildings, and a blue one found in tanks; all beautiful objects when displayed on paper.

Once when walking on the Wiltshire Downs I noticed curious bodies like half-empty grape skins. They were olive coloured, transparent, and gelatinous. I inquired of the natives regarding them; "they are Fallen Stars," said one, "Star Slough," said another. I tried to elicit a legend if I could not arrive at botanical information, but I failed even here. Nobody saw the stars fall, nor knew what the plant had to do with them, only it was called "Fallen Stars," or "Star Sloughs." I betook myself to my friend's well furnished library, and there ascertained my plant to be a Tremella, and consequently a fugus, and I placed it accordingly. But Greville and Hassak and Harvey now place it among Algæ, so my plant is removed to another volume (Nostoc Commune) Its cousin the River Nodularia is a pretty plant. Directed by Dr. Greville, I sought it in the stream traversing the glen between the Blackford and Braid hills. A charming hunting ground for flowerless plants is that valley. Secluded even from the sun's rays, funguses large and small luxuriating in the woods, and Algæ crowding both the land and the water, while the mosses outdo the foliage in luxuriance! There are weird stories, too, of the glen, and its long deserted Hermitage, and ghost and other legends hover cloudlike over the "puddock stools" and mossy encrustations. The Nodularia grows in dark green loose tufts, the stems are as thick as horse-hair, every here and there swelling into knots.

The Palmellaceæ group are composed of solid globules nestling in jelly. The Bloody Palmella (P. cruenta) grows on damp walls and rocks. I first noticed it on a wall at Kingston Deveril, in Wiltshire. It looks like fresh blood when wet, but losses its brilliancy, and turns powdery when dry. The green Palmella talso, found in Wiltshire, like deep green jelly among moss in thickets on the high Downs. The masses were shapeless, and from one to two inches broad (P. protuberans). The rose Palmella grows in minute globules as a parasite upon tree Lichens; I have it from the Braid valley, near Edinburgh. The Red Snow (Protococcus nivalis) is the near relation of the Palmella; it is formed of myriads of tiny grains, every grain a plant, and spreads over large surfaces, so that it used to be taken for a shower of bloody snow, and regarded as a phenomenon big with terror. The great speed of its reproduction was one cause of the marvel attending its appearance. Each grain contains numerous other grains; and, bursting, produces a large number of plants; which, in their turn, throw forth a crowd of offspring in an incredibly short period; thus in some few generations, each grain has literally become the "mother of millions." Snow seems to favour the increase of this Alga; and if a few plants have lain on the surface before the snow-nurse arrived, we can easily imagine how rapidly the minute grains might mingle with the light crystals, and give their own brilliant colouring to the mass. Sir John Ross describes the striking appearance produced by this snow-plant in Baffin's Bay, where he saw it covering miles in extent, and often penetrating to the depth of ten or twelve feet.

The succeeding group, the Diatomaceæ, are microscopic plants, with flinty coats; they are to the vegetable creation what zoophytes are to the animal; each weed consists of a group of cells, every cell a perfect plant, of a geometrical shape, but joined to its neighbour by one corner; independent, yet sharing the sap of the other subjects in its little kingdom.

A large family of plants still remain unplaced, because their position is hardly yet pronounced upon by the law-givers of botany. These are the Charas. By right of their brick red male organs, situated singly on the stem,

Linnæus placed them in the first class of Flowering Plants. Others have since counted them as fern allies; and their whorls of leaves have a certain resemblance to Horse-tails. Then, again, their immersed habit, and entirely cellular structure, seems to indicate Algæ as their natural companions; so let us place them here, for the present at any rate. These plants are formed of simple or compound articulated threads; they bear spores, as well as the brick red globules mentioned above; these germinate by the formation of a cell above the centre of the spore. As the plant grows, it assumes the form of a branching stem, with whorls of leaves at certain distances; the fructification appears in the axils of the whorls. The outside of these whorls is coated with carbonate of lime in most species. The common Chara first met my eye when a party of us were seeking for fresh water Mollusks in the Warwickshire canal, near Hawksborough. As the boys landed each hawl of weed, and after picking off the shells, flung it aside, I subjected it to a fresh inspection. I had secured two land-weeds which were new to me: perplexed myself afresh with the then undescribed plant of recent immigration, the Water Thyme; and was disentangling some Conferva threads, when the rough whorls enclosing the circle of red globules attracted my attention. I recogized it as a Chara from the description in the English botany. It was harsh to the touch, and of a whitish green (Chara Vulgaris). The shining Chara (C. translucens) I got from the sixty-two pools left by the receding waters of the Wye. In general form it resembles the common species, but it is smooth and transparent, apparently without a limy coating. There are several other species, but I have not found them. The most curious circumstance in the history of Chara is their circulation. The grains of coloured matter in each joint are disposed in two spiral bands, so as to leave an uncoloured space between. A circulation of the coloured mass takes place from below to the top of the cell, and then down again. If the joint is tied in the middle, then the fluid of each half circulates in the same way. This phenomenon may be observed in the shining Chara by the use of the microscope. It is said that the Malaria, hovering over the Pontine and other marshes, is caused by the odour of sulphuretted hydrogen emitted by Characeæ.

The fresh-water tribes of green Algæ are most numerous and pervasive. Professor Harvey thus writes of their habitats and uses :—" The Chlorosperms are more widely diffused than any other Algæ. A comparatively small number are found in the waters of the sea. A far larger portion inhabit fresh-water rivers, lakes and ponds, ditches, bog holes, the gutters of houses, and sewers; in fact, anywhere that fresh or unfresh water may lie; nor are they absent from the hot springs of volcanic regions; and are capable of vegetating wherever moisture and a moderate temperature prevail. Thus universally dispersed, they answer many a good purpose in the household of nature; and are specially useful in purifying the water in which they live. Unsightly as the green scum may be which they form on its surface, the growth is a renovating process, in which are consumed the deleterious matters and gases which stagnant water generally contains ; while, like all green plants, they pour into the atmosphere, during sunshine, oxygen prepared in their delicate tissues from the carbonic acid on which they feed."

1 Rose Baeomyces. 2 Red do. 3 Short stalked *same lichen* 4 Black G. 5 Goute G. 6 Greek Opegrapha. 7 Black O. 8 Variable O. 9 Submerged *Endocarpon* 10 Grey E. 11 Common Pertusaria. 12 Brown Opegrapha. 13 *Pleocarpon* 14 Horn O. 15 Birch O. 16 Starry O. 17 Speckled O. 18 Rusty Goblet *lichen*.

RAMBLES
IN
Search of Flowerless Plants.

CHAPTER XXI.

LICHENS.

"The living stains which Nature's hand alone,
Profuse of life, pours forth upon the stone ;
For ever growing ; where the common eye
Can but the bare and rocky bed descry ;
There Science loves to trace her tribes minute,
The juiceless foliage, and the tasteless fruit ;
There she perceives them round the surface creep,
And while they meet, their due distinction keep ;
Mix'd, but not blended, each its name retains ;
And these are Nature's ever-during stains !"
CRABBE.

LICHENS are plants coming next to the Mosses in botanical order, though differing widely from them in appearance. They are distinguished from seaweeds by the presence of minute green bodies (*gonidia*), lying generally in a layer between the upper and lower covering of the plant. The infant lichen first appears as a frail network upon the stone or bark, a layer of cells grows upon this, and then the gonidia are formed. Lichens are either *crustaceous* or *frondose*. Their fruit is of two kinds, the more perfect form being deposited in concave or convex shields, and lines; the less perfect in powdery warts. These sturdy plants seem strangely independent of the substance on which they grow. Some flourish on the hardest rocks, others prosper on healthy trees : they will bear all vicissitudes of weather ; for though they seem to dry up and die in the hot sunshine, yet the first rainy day enables them to expand again, and resume the business of their life. They have a wonderful power of retaining moisture, and also of collecting it; for if there is any damp in its neighbourhood, the lichen seems to attract it to itself. In dying they deposit a subtle acid, which wears away the rock on which they grew, and thus forms earth fit to nourish minute plants of higher organization. It was this capacity in the lichen to which the poet alludes, when describing the gradual rise and vegetation of a coral island—

" But soon the lichen fixeth there, and dying, diggeth its own grave,
And softening suns and splitting frosts crumble the reluctant surface."

But the progress of such vegetation must be at a rate inconceivably slow ; for, although the lichens attain maturity quickly, they are very long before beginning to decay. Mr. Berkeley states that he has noticed some for twenty-five years remaining in an unaltered state.

The first use of the lichen had long been familiar to me : but I was astonished to find many other uses for one or other of the tribe. As dyes, as food for animal and man, and as medicine, they have performed, and do perform, no mean part in vegetable economy. I was impatient to make personal acquaintance with them, and resolved to sally forth in search thereof immediately on my arrival at Hawkhurst.

A small plot of high ground, still waste, where the purple ling and dwarf furze flourish, was my first hunting ground. The plot in question rejoices in the name of "Starvegoose," in allusion, I suppose, to its barrenness. At noon, then, I found myself traversing the wooded road leading up from the village to this last plot of moorland, and quickly I transferred myself from the sandy road to the other side of the barrier gate. I took one glance around on that wide-spread landscape so essentially English; the hop-gardens were still mere forests of poles, and the corn-fields bore an emerald hue. Yet, though the rich woods were leafless and the orchards bare, the form of beauty was present in the variety of hill and hollow; and gray towers, rising now from woods or park-like fields, and now as land-marks on the seaward cliffs, pointed the eye upward to a clear blue sky, flecked with white clouds, where neither the form nor tint of beauty was absent. They pointed the eye—yes, and the heart too; for a soundless voice seemed to issue from them, mingling with the song of birds and the distant tinkle of the sheep-bells, saying, "Set your affections on things above."

A rough path traversed Starvegoose, and, seeing it coloured with various tints, I knelt to examine it. A greenish gray crust spread along the ground, sometimes thicker and sometimes thinner, so as to form an uneven surface; from this crust little bull-necked stems arose, surmounted by a rose-coloured top, resembling an uneven mushroom, except that it had no gills beneath. Here was certainly one of the *crusted* lichens; its true fruit (Apothecia) was well developed, and the form thereof showed it to be the rose-coloured mushroom lichen (Bæomyces roseus, *Plate XIII.*, *fig.* 1). I remember finding a plant resembling this at Braid Hermitage, near Edinburgh, when I was searching for fungi. It differed from the one in my hand in having a more decidedly green crust, and in the miniature mushrooms being more even in shape and of a brownish red hue. I felt glad that I had preserved the specimen; for it would make my collection better to have two species of Hooker's first genus of lichens. The Braid plant was the red mushroom-lichen (B. rufus, *Plate XIII.*, *fig.* 2). There is a brown species found on rocks and walls, and another, characterized by the thickness of its crust, which is peculiar to Ross-shire.

But time was passing away too fast to allow me to continue my reflections and my search together; so I hastened to collect all the beauties around me. There was a white-branched lichen of most elegant form, which I suspected to be the reindeer moss, and a lichen bearing little cups on its stem. Another, growing near it, had crimson knobs instead of cups, and in many places the stones and earth were covered with black swollen dots. These I placed in my case, and then betook myself to the woods, in search of further treasures.

Although the sun was shining so gloriously to-day, yet April showers had been falling heavily during the night, and for many days previous, so that the woods were very wet, and drops of rain still hung upon the tall weeds. I disentangled some decayed branches from under these weeds, hoping to find some lichens upon them. I was not disappointed. Numerous thread-shaped stems bore tiny heads, so that the end of the stick where they grew resembled a Lilliputian pin-cushion with the pins half drawn out. Upon applying my pocket lens I saw that these heads were goblet-shaped. Some had discharged their seeds, and were empty, and there was a thin blackish crust spread upon the stick. These peculiarities decided me in naming it the short-stalked goblet-lichen (Calicium curtum, *Plate XIII., fig.* 3). Upon the same dead branch I found another of the goblet-lichens; the stem was generally absent, and the cup broader, and of a reddish colour. This was the rusty goblet-lichen (Calicium ferrugineum, *fig.* 18).

Proceeding along the overgrown path I came to a rude bridge spanning the slow stream. Laying my hand on the rail, I perceived that it became stained as with soot. Again I produced my lens, and found that a lichen was growing thickly on the wood; this one was stemless, but still goblet-shaped. I remembered that the sooty powder was a feature in the black goblet-lichen (C. tympanellum, *Plate XIII., fig.* 4). In a lane near, I was afterwards tempted to search very closely by espying a patch of the Awl-leaved Earth moss on the bank, so I stooped very low, examining the ground inch by inch. A rotten stump grew in the hedge, and my curious eyes wandered around this, and into its dark cavities. There a beautiful sight greeted me,-a sulphur coloured stain, beset by stems and heads more delicate in form and more beautiful in colour than any goblet-lichen I had yet seen. The golden dust powdering crust, and stem, and head convinced me that it was the gold headed species (C. chrysocephalum, *fig.* 5).

Some months later, when spring had almost come round again, a party of us were wandering in the beautiful woods about Ross, in Herefordshire, making nosegays of primroses, and wild daffodils, and wood rushes, and verdant moss. We came to where the path passed between huge masses of old red sandstone rock, and paused to examine the miniature sward with which they were partially clothed. Here were the rich array of Liverworts which I have described in another place, covering the red sand plains and quartz hillocks with their minute, but elaborate lace-work. Behind them rose clusters of the Tree-line Feather moss, like sheltering woods, and closely crowded under their protection, nourished by a decaying fern stem, was a forest of wiry threads surrounded by heads like the most tiny beads. By very close inspection those with strong sight could detect a green colouring as a carpet for these miniature plants, the heads of some of which had burst, and being closely huddled together the spores had made a soft mat over the top of the stems; this was another species of Calicium (C. clavellum, *Plate XIII.*, *fig.* 5), and there are several more, but they have not rewarded our search.

The next family of the crustaceous lichens, as given by Hooker, is called *Arthonia*. The crust is thin and spreading, the fruit round and sessile. They form greyish stains, dotted with tiny specks of brown or black, upon living trees. Though not uncommon, I did not succeed in finding any of them on that occasion ; but on closely examining the bark of the trees, young and old, I did descry stains and markings of most wondrous form. Some seemed like miniature inscriptions in ancient characters, and some like interrupted lines, or irregular dots. There could be no doubt that these were the writing-lichens, and one in particular resembled so closely Greek characters that I named it the Greek writing-lichen (Opegrapha scripta, *fig.* 6.)

There was one growing on the bark of the Spanish chestnut, in which the short lines were waved and turned in different directions. The crust was buffish, and the long-shaped fruit was black. The distinguishing peculiarity in the fruit of these writing-lichens is the long shape and the depressed mark down the centre. The parasite of the Spanish chestnut was the Variable Writing-lichen (O. varia, *Plate XIII., fig.* 8) ; and the black Writing-lichen (O. atra, *Plate XIII., fig.* 9), with its close horizontal lines, decorated the bark of many young oaks. There are a great number of species in this family. On a subsequent occasion when exploring the grounds about Craig House near Edinburgh, we found the circular patches of the Brain Writing-lichen encrusting the walls (O. cerebrina, *Plate XIII., fig.* 12) ; and both there, and in the Highlands we found the Stone species (O. saxatilis, *Plate XIII., fig.* 14). The Starry, the Red, and the Birch Writing-lichens are parasites on trees, and grew about Hawkhurst, (*Plate XIV., figs.* 15, 16).

"These graphic lichens," says Dr. Murray, "inscribe in curious hieroglyphics, yet intelligible characters, on various barks, their worth or worthlessness. Some of these epiphytes are discovered on particular officinal barks, and not on others ; enabling us thus to distinguish them ; while peculiar lichens or fungi make their appearance, as soon as these officinal barks have lost their value; and thus stamp, as it were, the broad arrow of condemnation on them. These symbols of creation, therefore, when rightly interpreted, become a valuable key to practically useful knowledge.

Upon the bark of a tree in that charming Kentish wood, I descried another kind of lichen, with a thin grey crust, round black receptacles, and a tiny light dot in the centre. This was the gem-like Wart lichen (Verrucaria geminata) a member of the family named, because of their crust being covered with minute warts. A black stem covering the upper surface of a stone beside the brook, showed, through the lens, a similar structure ; and, after careful examination, I was satisfied that it was the black wart lichen (V. nigrescens).

Another species of this family I found subsequently. While staying in Swaledale a party of us rambled to the village of Healhaugh. Some went to visit the cottagers, who counted much on such occasional courtesies ; and the rest wandered in the fields in hopes of finding something to add to our botanical collection. A purling brook ran at the back of the village, hurrying from its source among the limestone rocks on the purple moor, yonder, to join the Swale, which was gliding serpent-like along the broad lap of the valley. This brook was fringed with Broaklime, and Water Figwort, and Watercresses; its clear stream showed every pebble in its bed, and many an algæ floating from them. These pebbles were, many of them, stained with black, own sisters to my Kentish friend; but some bore green stains, on which black receptacles were distinctly visible, and we hailed them as specimens of the submerged Wart lichen (V. submersa).

Next in order comes the group of internal-fruited lichens (Endocarpon.) They have no raised fruit, only dots in the upper covering of the leaf, these dots being the mouth of the sunken receptacles. We found a member of this family on a dripping rock near Richmond, in Yorkshire. The fronds were greyish brown, leathery in texture and lined with a darker shade, it was the greyish green Endocarpon (E. miniatum, *Plate XIII., fig.* 10). Some of our party afterwards found the same species on rocks of the same formation in the beautiful hills of Cheddar, Somersetshire.

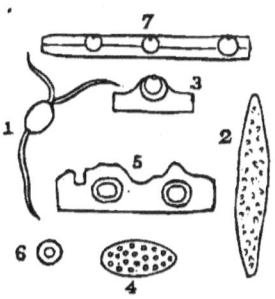

1. SPORE GERMINATING.
2. APOTHECIA OF OPEGRAPHA.
3. DO. OF VERRUCARIA.
4. DO. OF URCEOLARIA.
5. DO. OF PERTUSARIA.
6. DO. GONIDIUM.
7. SECTION OF ENDOCARPON WITH APOTHECIÆ.

This is the largest species in the family; there is an emerald species, and a dark grey, and an ash-coloured, and many others, growing on rocks or bark in various places, but I sought for them about Hawkhurst in vain.

Some fine old oaks standing at intervals along the

fields furnished me with good specimens of the Hole lichens, so called from the depressed points in the warts (Pertusaria). The common species (P. communis, *Plate XIII., fig.* 11), is like a circular grey patch, with warts crowded against one another.

Of the nearly allied family, the perforated lichens (Thelotrema), I could find no specimen.

There were no scattered-spot lichens to be found (Spiloma), though I remembered often to have seen the minute confluent receptacles of the wall species upon mortar. This family also contains many members, but they are too minute, and resemble each other too closely to draw much general interest.

In a wood, composed partly of firs, near the toll-gate, I found a pretty circular lichen; the margin was zoned and the inner part was sprinkled with white powder, while several shields were planted towards the centre. These were the receptacles, and boasted a torn border. This was the circular pustula lichen, the family name being the result of the swollen receptacles (Variolaria globulifera.) A tree near the edge of the wood bore large patches of the "inelegant pustula lichen" (V. agelæa); its grey crust with patches of white powder was so unassuming as to deserve the term inelegant. But these lichens, formerly erected into a separate family, Variolaria, are now ascertained to be only gonidial states of other lichens.

I now hastened my return, as the dinner hour could not be far distant; in this, however, I was deceived, for I had passed the off-shoot of the original village which is now called Highgate, when the unerring clock of the fine old church sounded the hour previous to our feeding time. This church had been restored since last I had held commune with the "men o' Kent," so I determined to spend the half hour now upon my hands in looking round its venerable walls. Here I soon found myself criticising the windward side of the tombstones with the aid of my pocket lens. There were patches of spangle lichen, both of the common and limestone species (Urceolaria calcarea, *Plate XIV.*, *fig.* 8), and the crab's-eye litchen, and several others. The spangle lichens are distinguished by the vase-like shape of their receptacles. As thus I sat, wearied with my long ramble, and thinking now of the varieties of living lichens, and now of the solemn dead around me, I felt a deep quietness steal over my spirit, and the poetic words of Ruskin recurred to my mind. Speaking of mosses and lichens, he says, "They will not be gathered, like the flowers, for chaplet or love token, but of these the wild bird will make its nest, and the tired child his pillow. And as the earth's first mercy, so are they its last gift to us. When all farther service is vain from plant and tree, the soft mosses and grey lichens take up their watch by the headstone. The woods, the blossoms, the gift-bearing grasses have done their parts for time, but these do service for ever! Trees for the builder's yard, corn for the granary, moss for the grave!"

As I drew near the fine old tiled dwelling of my friend, I rapidly glanced over the successes of my day. Of the ten first Hookerian families of lichens I had got specimens of eight, all being crustaceous except the internal-fruited and the leprous lichens. Two more crustaceous families remained to be studied on a future day; but I had at any rate got an idea of the form and arrangement of them, and the varied 'style of their receptacles. My ramble had been full of interest, and yet, certainly, these minute lichens were undoubtedly the least attractive of the order.

CHAPTER XXII.

LICHENS.

"While yet the forest trees
Were young upon the unviolated earth,
And yet the moss-stains on the rocks were new."
BRYANT.

"The lichens that so love to hide,
From those who have no eyes to see
God's beauteous work in things so wee."

THE bright spring days tempted us into the Yorkshire woods, and while we added mosses to our collection, and watched the early attempts of the fern crooks to uncoil themselves, we noticed various bright stains on rock and tree. Now the dark grey bark seemed powdered with sulphur, or brilliant violet, and now the frowning rock would be stained with soft patches of white or glaucous green. We surprised an eminent scholar and mathematician standing with folded arms, wrapt in admiration of a group of rocks thus coloured; he was a painfully shy man, and was generally known to avoid acquaintances in his rambles; but he advanced eagerly toward us, warmed out of his shyness by his wish to gain information, and pointing to the rocks, he exclaimed, "How beautiful the colouring is! tell me the cause of those rich tints?"

The stains were ascribable to four different species of Lepraria, a family where the spores are diffused among the powdery substance of the crust. They are common lichens, and being of such a low state of development, they are not accounted interesting members of their class.

All lichens are divided into two great classes, the *Crustaceous and frondose*, the thallus in the first division, consisting of a mere *crust* spread upon the rock or bark, and partaking all its inequalities; the latter having a leafy thallus, the substance outspread in lobed or branch-like fronds. The earlier groups of lichens belong to the Crust Division, with the exception of the Endocarpon, where the thallus expands in lobed fronds.

A very large family now comes in the regular order; that of Lecidea. Here we have a decided advance in the scale of development. The thallus or frond becomes somewhat leafy round the circumference, and the fruit cases (Apothecia) are salver-shaped, with a raised border of the same colour as the disk. The substance of the frond is still crustaceous, but it radiates in a regular form round the edge. The most remarkable member of this family is the Map Lecidea (L. geographica, *Plate XIV., fig.* 2), the thallus is bright yellow, but broken into unequal divisions, by black lines, resembling the divisions between counties in a map, while the black Apothecæ, dotted up and down, and varying in size, resemble the towns and villages there marked. Dr. Murray speaks of this lichen as imparting to the surface of rocks a very beautiful character. "Some of the rocks," he says, "around the Lake of the four Cantons in Switzerland, cannot be sufficiently admired on this account." This lichen affects trap rocks; we found it on such in the Isle of Arran.

The Rocks Lecidea (Petræa, *Plate XIV.*, *fig.* 1), is scarcely less attractive, its crust is white, and its Apothecia black and elongated, and clustered in lines diverging from a centre. We found it on a remarkable excursion to the Cheese Wring, in Cornwall, where huge masses of granite rock are heaped in a circle, bearing evidence of druidical origin. Great numbers of quarrymen were engaged in excavating the granite in the neighbourhood, and again and again as we searched for flowers and lichens, the ground beneath us was shaken as with the shock of an earthquake, owing to the blasting necessary to detach the granite blocks.

On mossy trunks, both in Yorkshire and Kentish woods, we have often found the Yellow Lecidea (L. lutea, *Plate XIV.*, *fig.* 5). Here the crust is pale yellow, and the Apothecia darker yellow. Black dots of varying size, surrounded by a scarcely perceptible grey crust, formed patches on the limestone rocks in beautiful Swaledale (L. atrata, *Plate XIV.*,*fig.* 4), and others seemed to have eaten into the stone, and buried themselves each in a little nest of its own excavation (L. immeria, *fig.* 3). Mope's Den, a beautiful wood in the Bedgebury Estate, near Hawkhurst, furnished us with a green Lecidea (L. parasema, *Plate XIV.*, *fig.* 6), which grew on the trees, and bore black receptacles; whilst stones lying on damp places on Starvegoose heath supplied the Bog

Lecidea (L. uliginosa, *Plate XIV.*, *fig.* 7), a collection of larger black apothecia on a thin grey crust. This family numbers above sixty English members, varying in colour to every shade of grey and green, and with brown, yellow, grey, and black apothecia. They live indiscriminately on bark and stone.

A small family called Urceolaria, forms extensive patches of crust on rocks and walls, especially in limestone districts. The Calcareous urceolaria is grey, with grey chequered depressed lines, and small grey apothecia (U. calcarea, *Plate XIV.*, *fig.* 8), we found it abundantly in the Yorkshire dales.

A very large family comes next in which there are some important members. One, in particular (Lecanora esculenta), presents the strange anomaly of a free lichen, unattached to any habitat of wood or stone, and drawing its entire subsistence from the air. The Edible Lecanora is an Asiatic species, the apothecia are found scattered on the ground, some as large as a walnut, and are used in conjunction with wheat, as food by the Tartars. In the English members of the family the frond is still crustaceous, but increasingly foliated at the margin; the apothecia have generally borders of a different colour to the disk, marking their position as belonging to the Cænothalami, those lichens whose apothecia is formed *partly* of the substance of the thallus. There are groups of rocks in Swaledale which afford first rate opportunities of studying the Lecanora family. Recently taken into the enclosure surrounding a country house, convenient paths now intersect the wilderness; and although there is still ample opportunity for hiding, or to some extent losing yourself, and for breaking your legs among deep fissures in the grotesque rock masses, yet their treasures are brought within much more easy reach of the adventurer, and you may chance to find a rough board nailed across two blocks of limestone of a sufficiently similar height to afford a safe resting-place, near to a mass of weather-worn rock, encrusted with the Crab's eye lichen and the Cudbear. The Crab's eye lichen (Lecanora Parella, *Plate XIV., fig.* 10), forms an exception to the rule recently laid down, of the disk of the apothecia being different in colour to the crust. In this instance the colour is the same. This lichen abounds on rocks and walls, covering the weather side of gravestones with circular grey warted crust, flourishing in the graveyards about Hawkhurst, and in those about Edinburgh, and decorating the rocks at Tunbridge Wells, granite rocks in Cornwall, those of limestone in the north of Yorkshire, and the trap and puddingstone of the Highlands of Scotland.

We should select the Cudbear as the head of the family (Lecanora tartarea, *Plate XIV., fig.* 11) in Britain, seeing that it has afforded an important article of commerce. It takes its English name from that of Dr. Cuthbert Gordon, who first discovered the presence of the colouring pigment in it, and employed it largely as a dye. The gathering of this lichen used to furnish the means of living to great numbers of the poor in Wales and in the Highlands; they scraped the rocks with an iron hoop, and sold the scrapings at a good price; each rock yielded a crop once in five years. A foreign species, growing in the Canaries, is now preferred to our English Cudbear, so that it is no more gathered for the market. In former days it used to yield the best scarlet dye; and, as such, it was plentifully employed in the manufacture of the cloaks then so much in vogue. It did good service to the country at the end of the last century; for when the French troops landed at Fishguard, and were opposed only by a very inadequate body of yeomanry, the sight of a number of Welsh women in the distance, mounted on hill ponies, and wearing their high hats and Cudbear-dyed cloaks, bound, I suppose, for the nearest market, struck them with panic, because they mistook the cloaks for the "red coats of the regulars," and they surrendered themselves to Lord Cawdor and his yeomanry without striking a blow. The Cudbear grows abundantly on these Swaledale rocks, its ochre shields crowding so close together that they shoulder each other out of shape; those that have the good fortune to get more elbow-room grow as large as a threepenny piece. The crust is of a whitish gray, like that of the Crab's-eye lichen.

The black-shielded Lecanora is there too (L. atra). Its crust is whitish, rugged, and cracked, its shields oval, but often squeezed, so as to be quite narrow, the disks full black. For our specimens of the Red-spangled Lecanora (L. ventosa) we are indebted to the rocks about Arthur's Seat. It is a handsome species, with ochre crust and dark red shields. It yields a purple dye: indeed, many lichens of this family yield dyes of purple, red, or brown; and Hellot gives a simple recipe for detecting the colouring principle:—"Put half an ounce of the plant into a glass, and moisten it with equal parts of lime, water, and spirits of sal-ammoniac; tie a wet bladder close over the top, and let it stand three or four days. If any colour is likely to be obtained, it will then be evident in the liquid."

When climbing to the topmost group, we overlook the beautiful valley of the Swale and the crowd of gray hills stretching beyond it on every side till their outline becomes so faint as to be hardly distinguishable from the horizon, and still behold the Crab's-eye and Cudbear peopling the commanding rocks, we think of Darwin's description of the plant :—

"Retiring lichen climbs the topmost stone,
And drinks the aerial solitude alone."

One February, when exploring the banks of the Looe river, near Looe in Cornwall, we came to an old wall sometimes formed of sandstone rock, sometimes built into it. Here we first found members of the Psoræ, or Scurf lichen family. One composed of light green scales lined with white, and bearing orange shield-like receptacles, was the Brown Scurf lichen (Psoræ globulosa, *Plate XIV.*, *fig.* 13).

Near it was a black and gray-speckled lichen, of the same form, the so-called Black and Blue Scurf lichen (P. ceruleo nigricans, *Plate XIV.*, *fig.* 14.) Some small orange specks on the ground showed similar structure under the magnifier, and thus declared itself to be the Black and Red Scurf lichen (P atro-rufa). There is the raspberry-fruited and the staircase species ; but we have not found either.

The next family is that of the Squamariæ or Scale lichens. The Shield lichens are the last of the crusted group; those that follow are becoming more and more *leafy*. In this family the frond is starry, and the receptacles bordered and sessile. The handsome Orange Wall-scale lichen was half-covering the rock in question, and we procured some good specimens (Squamariæ murorum, *Plate XIV., fig.* 15). This is a very common but very handsome lichen. It covers rocks near the shore in Cornwall, trees and roofs in Kent, and old walls in Yorkshire, and one near Melbeck's Parsonage, in Swaledale, is always glowing with its amber fronds, as if in midsummer sunshine.

A duty walk, a regular "constitutional" among the chalk downs, one January day provided me with specimens of the Stone Squamariæ (S. saxicola, *Plate XIV., fig.* 16). They were greenish yellow stars, with a cluster of green apotheciæ in the centre, and studded the flints which lay here and there upon the short herbage, as they had cropped out from the chalk-land beneath. The Circular Squamariæ (S. circinnata, *Plate XIV., fig.* 17) is common in Yorkshire. It is grayish brown, darkest in the centre, and much furrowed as with veins. It radiates towards the margin, which is quite lobed and leafy, and substantiates its claim along with others of its family to be considered a step towards a frondose lichen. In this group the thallus is closely attached to the rock as far as its uttermost margin.

The Candle lichen (S. candelaria), used in Sweden to impart a yellow stain to the candles employed on festive occasions) is a member of this family, and is the only one among the many British species to which any use is assigned.

There are numerous other species of this family, but only these have rewarded our research.

Of the genus Placodium we have never succeeded in finding a specimen.

The Parmelia group contains some handsome and showy species, the theme alike of poets and lovers of nature. The fronds are scaly in the middle, and attached by fibres or a narrow vase to the substance on which they grow, but they are free for the remainder of their breadth, lobed and veined, and have a full claim to rank as frondose lichens. Upon poles, near the Dever, in Wiltshire, we found the Sulphur Parmelia (P. caperata, *Plate XIV., fig.* 17). It is a handsome spreading lichen, powdery and pale yellow above, dark brown and hairy beneath. Brown is the colour of the apothecia, but our specimens had no fruit upon them.

The Crotal or Crostal of Ireland and Scotland is a member of this family (P. omphalodes, *Plate XIV., fig.* 18). I first saw it upon rocks in the "black country," bordering Lord Breadalbane's deer park in the Western Highlands. It is a wonderful district; and as the coach wound slowly up the steep road, constituting, as the driver informed us, "the highest travelling ground in Scotland," we noticed dark bronze patches on the surface of the gray rocks. We had passed the pine groves, and the herds of shy deer and the shooting lodge beyond, and had come into a land of peat bogs, black, as if drained from coal-pits, and brown heath, and rugged rocks,—no green in moss or grass, no flowers, no gay colouring. Here it was that the Crotal was flourishing, its leafy patches forming glossy stars of brown madder. It is used as a brown dye for the home-spun fabrics of the district; and Walker declares it to be the most indestructible of colours. The apotheciæ, which it freely bears, have coal black disks.

Rocks in Swaledale, as well as about Oban and Callander, furnished us with the Rock Parmelia (P. saxatilis, *Plate XIV., fig.* 19). This is a pretty species when closely examined, somewhat resembling the Circular Squamariæ in its dark centre, rutted substance, and lighter margin; but it has the family characteristic of only being attached to the stone by fibres, and so is easily distinguishable.

Another Cornish ramble was to the Chough rock, a large flat-topped cliff overlooking miles and miles of blue sea, along which the broad Atlantic waves were rolling; while vessels bound from Plymouth to Falmouth were making their noiseless way across the blue expanse. Here, upon the shelves of the rock, were patches of the sunburnt Parmelia (P. aquila, *fig.* 20); its glossy brown fronds, and chocolate-disked apothecia, identifying it as that species. We listened to the sweet harmony of the waves breaking on the shore, and saw them deposit their successive burdens of gay weeds. This lured us to the sands, and we climbed down by a steep path to fill a bag with seaweeds and shells, and pick up any stray lichen that chance might send for our basket.

And here the golden Parmelia was flourishing in rich luxuriance, its tile-like fronds overlapping one another, and bearing such abundance of full orange shields that it seemed as if sea-air were its favourite enjoyment (*Plate XV., fig.* 15). It is called the wall Palmelia, and rightly, for it adorns walls and tiled roofs as freely as the wall Squamaria does. This lichen also affects trees, adhering to the rutted bark, or encircling the branches of oak or thorn.

> "The yellow moss in scaly rings
> Creeps round the hawthorn's prickly bough."
> WORDSWORTH.

> "All the same, ever the same, this outward face of things,
> Time but toucheth it gently, little the change that it brings;
> Here where we sat together, spreadeth the self-same tree,
> Carved and matted the branches, just as they used to be.
> Even the rich-toned lichen keepeth its place and form,
> Mellowing the old gray oak-bark, tinting it sunset warm."
> CHAMBERS'S JOURNAL.

A pretty pale species, the Pearly Parmelia (P. perlata,) was sent to us from Malvern; Lees, the graceful writer upon Malvern botany, says, "The rocks are hoary and silver grey from an extensive spread of Parmelia Saxatilis, and Perlata;" and we can well imagine the glossy fronds of the pearly Parmelia giving a charming variety to the dark silurian rocks of that beautiful district.

The nineteenth family of lichens is that of the dotted group Sticta, so called from the depressed dots on the under surface of the thallus. The fronds are large, lobed, downy beneath, and quite free, except at the base.

We have all heard of the "Lungs of the oak" (Sticta Pulmonaria, *Plate XIV., fig.* 9), but none who have not seen it can imagine its striking appearance. Among the party on the coach as we passed through the Trosachs, it created great interest, all wondered what the "strange leafy thing" covering the boles of elms and oaks for several feet from the ground could be. The patches were more than a foot broad; sometimes adhering closely to the bark, but oftener with several of the large lobes turned back, shewing the veined and pitted under surface, studded with hollows and grey down. In the young plants the colour was light green; but in more advanced age it was olive brown; and the actually aged ones were grey. In this family the Apotheciæ are very small, and are situated on the under side of the frond. We made an excuse to save the horses, and walked up the hills, taking that opportunity of securing large sheets of the kingly lichen, our new and admired acquaintance. This used to be given as a remedy for consumption in former days, either on account of its possessing, in some small degree, the bitter stomachic principle, which has rendered the Iceland moss so deservedly esteemed, or upon the less reasonable doctrine of initials, where the outward form was held to be the sign of a hidden use, and the lung-like shape of the lobes, with its pitted cavities, suggested them as remedies for lung disease. In these days such logic appears the very perfection of absurdity; but in the minds of many deep thinkers there is a strong persuasion that we are but on the surface of botanical knowledge, and that in the secrets of science yet unfathomed, an analogy between form and application will one day be found, which will at once lay open to every observant mind the qualities of each plant upon which his thoughtful gaze fixes.

We did not find the pitted Sticta, but a specimen of it was given to us by one who had the opportunity of exploring the highlands much more thoroughly than we were able to do. The pitted species (S. scrobiculata, *Plate XIV., fig.* 24), is a very handsome lichen, of a dull green externally, and ornamented by powdery warts containing clusters of gonidia; the under side is dark and hairy, with deep pits honeycombing its surface. The yellow Sticta (S. crocata) is a magnificent species, olive and veined with yellow, and with lemon spots on its under surface. It grows in Ireland, and is very rare. The broad leaved Sticta is also an Irish species; and, like the yellow, is very scarce. The family contains several other members, but the two first mentioned are the only ones that grace our collection.

CHAPTER XXIII.

LICHENS.

" Where o'er the jutting rocks soft mosses creep,
Or coloured lichens with slow oozing weep."
COLERIDGE.

T was a great fair in the good town of Looe, and as botanical specimens were not among the commodities to be procured there, we took biscuits in our pockets, and set off for a long excursion, which should occupy great part of the short February day. We crossed the bridge, and took to the west side of the river, following its course seaward, and rounding the shoulder of the cliff shutting in the harbour, and then making for the rocks along the coast, called Hannaford.

From the steep path the scene was very striking. An open space close to the beach on the east side was filled with cattle, while the few narrow streets adjoining were crowded with farmers and drovers. Along the roads various groups of young beasts, which had already found purchasers, were being driven, and every ship in the harbour displayed its flag in honour of the important occasion. Away to sea lay the Looe Island, girt with rocks, between which and the shore the passage is so shallow that it is not safe for schooners, except when the tide is

1 Great Colrana. 2 Crisp C. 3 Green Socket licht. 4 ...
5 Burnt G. 6 Fleecy G. 7 Iceland Moss. 8 Snow ...
10 Fucus like Roccella. 11 Hairy Borrera. 12 Dwarf B. 13 Branny B. 14
Dog Lichen. 15 Wall Parmelia.

nearly at the height; and many an unwary vessel sticks fast there. A zostera bed marks the low tide on the mainland, and the crop of fucuses and "dead men's ropes" is most flourishing. As soon as I could gain the shore, I began climbing from rock to rock towards the mouth of the river. The ground beyond high-water mark was spongy, and the flat tops of many of the rocks were grassed over. Here we found our old friend, the Crab's-eye lichen, in abundance, accompanied by its faithful ally, the Black-shield lichen. The Yellow Scaly lichen and the Wall Parmelia were liberal of their orange shields and spreading fronds, and they coloured the grey rocks brilliantly. On one of the grassy flats I found what seemed like roughly-shapen, thickish leaves of green jelly, the under part paler than the upper. I could find no receptacles; but from the peculiar appearance of the plant, and its resemblance to the common nostoc, or "star-slough," I felt sure it was the Shaking Jelly lichen (Collema tremelloides). This was a member of the lichen order next succeeding to the Sticta, and called Collema from its glutinous structure. The plants of this group have round apothecia, bordered, the disk coloured with brown; their whole structure is gelatinous.

• On the steep wet banks beside the deep cutting forming the junction road near Hawkhurst, we had once found the gelatinous olive fronds of the Great Jelly lichen (C. granulatum, *Plate XV., fig.* 1). It was of a darker colour than the Hannaford species, but more regular in shape: it had no shields. Walls about Brixton Deverill in Wiltshire furnished us with abundance of the Crisp Jelly lichen (C. crispum, *Plate XV., fig.* 2). It grew in crevices between the stones in the wall, its thick semi-transparent dark olive fronds crowded together, and bearing numbers of chestnut apotheciæ, shield-shaped, and often so crowded together as to become confluent. But these were bygones on the day of our Hannaford expedition, and we pressed on from the rocks, which gradually became lower, and destitute of samphire, and wild beet and thrift—only now clothed with seaweed, showing that they were submerged at high tide; and we turned our attention to the sand banks, down which the water oozed from the upper land, encouraging the growth of many a moss and lichen. Here was a plot of gelatinous-looking shields, of an orange brown, bordered with the same colour; the fronds to which they were attached were almost covered by the receptacles, and were of a spongy nature, so that we decided this to be the Sponge Jelly lichen (C. spongiosum). Near it was a rough black stain, which under the magnifier showed many little gelatinous branches, with shields interspersed. It was very minute, but we were able at once to pronounce it the Dwarf Jelly lichen (C. subtile). Little orange shields scattered on the firm sand, their crust being scarcely perceptible, announced the presence of the Red and Black Scurf lichen; and the Ground Lecidea was there too, so that the barren sand-bank boasted its four lichens.

There are many more species of the Collema group, but they are all marked by the same characteristics, but these were all which we have been able to collect.

Leaving the shore we crossed some fields which brought us to a farm called Port Looe, and we pressed onwards along the lane leading from thence towards the

Trelawny woods. Both in the hedge-banks and among the richest moss in the woods, the Large Dog lichen displayed its leaf-like fronds. This plant measures several inches, and its tinted fronds, varying from grey to olive, look very beautiful beside the verdant moss. At first sight it gives one the idea of a torn kid glove; but the white underside is closely set with prickle-shaped hairs. The receptacles are orange, and they are freely disposed along the lobes of the frond; the border is the same colour as the disk.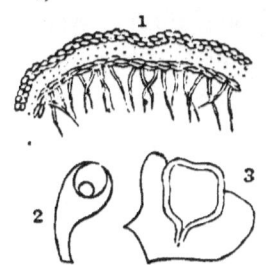

1. SECTION OF DOG LICHEN, SHOWING UPPER AND UNDER BARK AND MEDULLARY LAYER, WITH GONIDIA.
2. BRANCH OF GLOBE LICHEN.
3. APOTHECIÆ OF NEPHROME.

This, then, is the Dog lichen (Peltidea caninea, Plate XV., fig. 14), formerly believed to be an antidote to hydrophobia; and certainly it would be difficult to overpraise its beauty, whether it may be able to boast of any use or not. In gathering some of the handsome fronds of this lichen, we could not avoid tearing them. This afforded the opportunity for noticing its structure; and we sat down on the bole of a fallen tree to discuss our biscuits and examine our torn frond, and apply our Codington lens to ascertain the arrangement of the bark and gonidia. Even with the naked eye we could discern the upper covering or bark, the middle or medullary layer, and the under coat with its clinging fibres. We knew that the fruit, whether in its perfect form of shields, or its secondary development of powdery warts, took its rise from the medullary layer, the seat of the gonidia. By subjecting the section to the microscope afterwards, we saw the closely packed cells of the bark, the more loosely-collected and larger cells of the medullary layer containing the gonidia, and the elongated cells which form the fibrous under covering. This is a very common lichen, and grows as luxuriantly in the Highlands of Scotland and the dales of Yorkshire as on this southern peninsula, Cornwall. It is as beautiful as it is common, its leafy fronds, whether dry and gray, or moist and olive, and its plentiful chestnut shields, forming an attractive object among mosses, round tree roots, or in the Alpine pasture and the sheltered grave-yard. A pretty poem, which appeared some years since in Chambers' Journal, must have been suggested by this lichen :—

"Ye dainty mosses, lichens grey,
  Laid cheek on cheek in tender fold,
  Each with a soft smile day by day
    Returning to the mould.

"Brown leaves that with aerial grace,
  Slip from the branch like birds a-wing,
  Each leaving in the appointed place,
    Its bud of future spring.

"If we, God's sentient creatures knew
  But half your faith in our decay,
  We should not tremble as we do,
    When He calls clay to clay.

"But with an equal patience sweet
  We should put off this mortal gear,
  In whatsoe'er new form is meet,
    Content to re-appear.

> " Knowing each germ of life He gives,
> Must have in Him its source and rise ;
> Being that of His being lives
> May change, but never dies.
>
> " Ye dead leaves dropping soft and slow,
> Ye mosses green and lichens fair,
> Go to your graves, as I will go,
> For God is also there."

The genus Solorina contains but two species ; its characteristics are the orbicular apotheciæ and woolly veins. The Saffron Socket-lichen (Solorina crocea), is yellow or greenish, and its sunken apotheciæ are chestnut. It grows on ground at the top of mountains. We have none of us found it, only learned its characteristics from Hooker's description.

Once I had the good fortune to find the green Socket-lichen (Solorina saccata, *Plate XV., fig. 3*). We were wandering on the Yorkshire moors, with the intention of seeing a smelting mill, and collecting anything that could be found. We had left the broad valley, and with it all signs of verdure, and had followed the miner's path till we reached first the mill, and then the mouth of the mine. Long lines of carts drawn by a horse along the tramway brought lead ore from under the gloomy archway in the hillside, and deposited their burdens in heaps outside the mill. It was a July day, and how the sparkling metal glittered in the sun ! What a comfort it seemed to the heavy mill wheels as they laboured round and round, crushing the ore, that water was for ever dripping over them ! And with what an unearthly glare did the molten lead bubble and gurgle from the oven's mouth as the men drew it off into the prepared mould, while the refuse or " slay" rolled away in a crackling stream of livid fire, smoking and growing black as it cooled in the comparatively cold air of the mid-summer atmosphere. All this we watched with dutiful observation. We had come to see the process of smelting lead, and we must make the best of our opportunities; but when we had done our duty, and received specimens of the ore, we proceeded onwards to find a place beyond the sound of the mill, where we might gain shade and water, and a snug place to eat our luncheon.

But this desire was not so easy of attainment. All the rocks forming the back bone and ribs of those extensive moors are traversed with veins of lead ore, and the farther we went the more wholly did we seem lost among brown moor and grey crag, the only variation being the chimneys which ever and anon indicated that a branch of the numerous mines was excavated underneath. In time, however, the banks of the little stream to our left grew deeper and deeper, and we descended and followed its course.

> "Hey the green ribbon ; we kneeled beside it,
> We parted the grasses dewy and sheen ;
> Drop over drop there filtered and slided
> A tiny bright beak that glittered between.
>
> "Tinkle, tinkle, sweetly it sung to us,
> Light was our talk as of faëry bells—
> Faëry wedding bells sweetly rung to us
> Down in their fortunate parallels."

Onward we went, till we came to one of the loveliest nooks I have ever beheld. Our tiny brook must have been capable of great things in the times of winter floods, for its waters had scooped a perfect circus in the limestone rock, several yards across, and twenty feet in depth. Now the water oozes and trickles over the summit, and through the crevices, but the masses of stone lying in the bed revealed the force with which its waters could rush when snows are melting, or December rains falling.

It was in one of the crevices, sparkling with the spray of a neighbouring rivulet, that I espied the full green fronds and hollow black apothecia of the Green Socket lichen, and I enjoyed the refreshing meal which my friends were preparing, and the welcome shade of the brook's own drawing room all the more for the treasure that I had secreted in my basket. The Cudbear, and Crab's Eye were there too, and some of the Squamarias and Parmelias, and Apple moss and Feather mosses in abundance, and Wall-rue, Black-stalked Spleenwort, and Downy Hawkweed, but none had any charm in my eyes compared with that of the Green Socket lichen.

The Nephroma group, characterised by the kidney shaped lobes on which the chestnut-coloured apothecia are situated contains only two species, one inhabiting hilly rocky places, and the other quarries ; we have not been able to find either of them.

A curious group of lichens succeeds that of Nephroma. The Gyrophoræ have large round apothecia, in which the spores are arranged spirally, or in a line wound round and round on the surface of the flat shield. The Many-leaved Gyrophora (G. polyphylla, *Plate XV., fig.* 4), was lifting its leaf-like transparent olive fronds among the moss in damp places, beside the rocky path beyond the

Chough Rock. Here the ground was oozy, and though a great height above the sea, tiny streamlets were issuing here and there, which as the path descended, accompanied its slope, and on reaching the low ground joined a brook that watered the valley of Playday. There was no fruit on any of the fronds that we gathered, but we had the good fortune to find an apothecia on a plant of another species of Gyrophora, which we found adhering closely to the rock. This was a large circular sooty patch of wrinkled foliage, it was dry and very brittle, and as we detached it from the stone with our large knife, it split and cracked in various directions. This was the Burnt Gyrophora (G. deusta, *Plate XV., fig.* 5) and when wet, it was thin, transparent, and of a dark olive. A Scotch friend furnished us with a specimen of the handsome Fleecy Gyrophora (G. pellita, *Plate XV., fig.* 6), so called from the dense black hairs which cover the under surface, and which, as the broad lobes turn over on the plane surface of the green frond, gives the appearance of fur trimming to a lady's cloak. This lichen grew on the rocks on Lord Breadalbane's Park, but it was where we could not descend from the coach to obtain specimens.

Closely allied with the Gyrophora group is that of Umbilicaria, containing only one species, the famous Tripe de Roche (U. pustulata) so valuable as affording sustenance to Canadian Hunters. It is beautiful in appearance, greenish grey, having raised warts on its surface, the colour becoming darker towards the margin, which is fringed with black hairs; the under side is brown, and depressed in pits just where the upper surface is blistered. Dr. Richardson and his party were sustained for a long time by this mere leaf-like plant, when they were exploring the northern snows in pursuit of scientific objects.

An excursion among the Pentland hills furnished us with specimens of the most important members of the succeeding group, that of Cetraria. The Iceland moss is familiar to us all, as contained, dried, in glass jars, in the chemists' shops, and to such of us who have been accustomed to attend on invalids, the preparation of strong jelly from this lichen is equally familiar. It is a very important member of the lichen clan, not only for its medicinal but for its nutritive qualities. The Icelanders make a savoury dish of it, beating it with milk, and baking it in cakes. Henderson states that the porridge made from it is the most palatable food to be had in Iceland. Dr. Johnstone in his " Flora of Berwick upon Tweed" asserts that it is used in the manufacture of ship biscuits, preserving them from worms. But familiar as the lichen was to us as an article of commerce, it was a new treasure when growing in upright olive-brown tufts, each frond paler on the under side, and with a fringe of dark hairs (C. islandica, *Plate XV., fig.* 7). It grew under the shelter of sturdy furze and tufts of heath, and on the high Scotch mountains. Subalpine moors in Scotland are the only British habitat of this lichen, but it grows more luxuriantly and plentifully in high latitudes. The Snow Cetraria is a very pretty species (C. nivalis, *fig.* 8). It is of a very pale buff, almost ivory coloured, and becoming a full sulphur at the base; it grows upright, and its fronds are narrow and much divided. Our specimens were sent us from the Scotch mountains. The Glaucous cetraria (C. glauca, *Plate XV., fig.* 9) is common on the boles of trees everywhere. It grows horizontally at first, then lifts the lobes of its broad fronds almost perpendicularly, so as to show the dark lining in strong contrast with the glossy glaucous hue of the upper surface. We have found it in Wiltshire, Kent, Herefordshire, Yorkshire, and in both the lowlands and highlands of Scotland, but never got a specimen bearing fruit. The chestnut apothecia should be found near the margin of the frond, but we have sought them in vain.

Next in order comes the Orchil group, important as furnishing the species best adapted for dyeing. The Roccella Tinctoria, or Orchil of commerce, was once a source of wealth to the inhabitants of Cornwall and Jersey, where it grows pretty freely, and from it was procured the valuable colouring pigment; but it is now found that this pigment, as well as that extracted from the Cudbear, can be obtained in greater quantity and better quality from the allied lichens imported from the Canary Islands. The fronds of this lichen are rounded and upright. We found poor dwarfed specimens on the rocks about Looe, but the flat dark apothecia were nowhere present.

But the Fucus-like Roceella we found growing freely on those sea-ward rocks. It is much admired for its pale tint, its fronds are branched and flat, it is a pretty species (R. fuciformis, *Plate XV., fig.* 10). The Borrera group, so called from the botanist Borrer, who took great pains in studying them, is characterized by the branched and torn fronds being channelled beneath, and generally fringed around the margin; the apothecia are chestnut and shield-shaped. The fringed Borrera (B. ciliaris, *Plate XV., fig.* 11) is frequently found, and we have specimens from various localities, but the largest fronds and deepest fringe is on plants which we gathered off some stately elms in the glebe fields at Brixton Deverill, in Wiltshire. The thallus is greenish grey, the green tint predominating when moist, and the grey when dry, a brown tinge mingles with the grey on the under side, and the fringe bears the same tint. The chestnut apothecia are borne in abundance, and the curled lobes laid one over another, tier beyond tier, spreading into patches half a foot wide, are highly decorative to the Elm Bark. The Dwarf Borrera (B. tenalla, *Plate XV., fig.* 12) is a frequent parasite on thorn bushes. Its little fringed fronds, grey, and lined like those of its big brother, were encircling the rugged elbows of the Blackthorn bushes on the Chough rock, overlooking the Cornish sea; they clothed the Hawthorns in the Goblin Combe, Somersetshire, and endowed the thorny thickets on the Wiltshire Downs with hoary foliage as in the shrubs described by Wordsworth:—

> "Like rock or stone they are o'ergrown
> With lichens to the very top,
> And hung with heavy tufts of moss,
> A melancholy crop."

The Branny borrera (B. furfuracea, *Plate XV., fig.* 13) was growing on the boles of trees in the highlands. Its colour is rather brown than grey, and its narrower fronds are lined with black; a mealiness on the upper surface procures for the plant the specific name of "branny."

It is a handsome species, growing in extensive patches; and we were grateful to the wild woodland above Bracklyn Bridge for furnishing us so great a treasure. Tourists launch into well-deserved praises of that brawling stream with its magnificent rocks, and dashing waterfalls, and bending trees; but to us there was an added beauty, an extra claim on admiration, in the abundant lichens thrown together in wild profusion, hanging, as silver foliage, from the aged branches, and lapping one over another like tiling among the tall mosses. There, where everlasting moisture hangs, the mosses and lichens grow to double their usual size; the rocks are studded with crustaceous lichens—Lecidia, Lecanora, and Urceolaria—and draped with pendant masses of interwoven feather moss and oak lungs, fork moss and dog lichen, while hundreds of Yellow chanterelles, Fly agaries, and pink Russulæ spread their domes over the verdant tapestry.

1 Evernia. 2 Funale Ramalina 3 Ash R. 4 Rock R. 5 Old man's hair 6 Woolly Horn lichen. 7 Brittle Globe lichen. 8 Reindeer moss. 9 Branched Stereocaulon. 10 Common cup lichen. 11 Scaly C. 12 Coral C. 13 Finger C. 14 Pleurant C. 15 Compressed Globe lichen. 16 Hairy Usnea.

CHAPTER XXIV.

LICHENS.

"When in the grass sweet voices talk,
And strains of tiny music swell
From every moss-cup of the rock,
From every nameless blossom's well."—BRYANT.

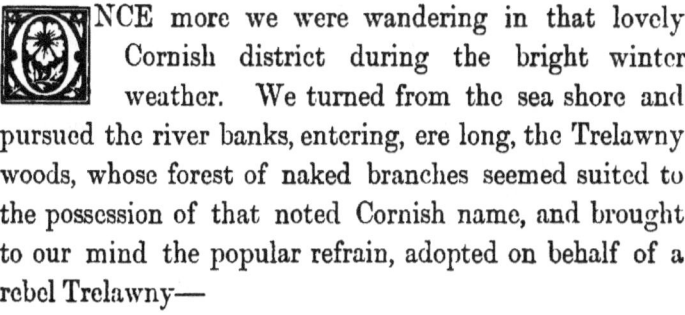NCE more we were wandering in that lovely Cornish district during the bright winter weather. We turned from the sea shore and pursued the river banks, entering, ere long, the Trelawny woods, whose forest of naked branches seemed suited to the possession of that noted Cornish name, and brought to our mind the popular refrain, adopted on behalf of a rebel Trelawny—

"And shall Trelawny die!
And must Trelawny die!
Then forty thousand Cornish men
Shall know the reason why."

But it is doing an injustice to the crowded oaks and elms of those Trelawny woods to call their branches "naked," for they were draped with lichens, just such as Bryant describes in the Canadian forests—

" Here are old trees, old oaks, and gnarled pines,
That stream with grey green mosses."

And again, when portraying patriarchal hunters, he calls them

"Less aged
Than the hoary trees and rocks
Around them."

Lifting a fallen branch from the ground we proceeded to examine the crop of lichen that it bore. There were scales of the orange Parmelia, patches of the Fringed and Dwarf Borrera, scars of writing lichen and Lecidea, but overshadowing all these, and absorbing the chief attention, were forked branches of whitish grey, covered with powdery warts, and bearing shields, with ochre-coloured disks and borders the same colour as the frond. This was the Hoary Evernia (E. prunastri, *Plate XVI., fig.* 1), one of the commonest of branched frondose lichens. It was growing here in tufts, by hundreds, on each aged tree, and we have found it as abundantly in Yorkshire, Herefordshire, and Kent. This lichen has a remarkable property of retaining odours, and was formerly much used in the preparation of perfume powder. Evelyn says, "Of the very moss of the oak, that which is white composes the choicest cypress powder, which is esteemed good for the head; but impostors commonly vend other mosses under that name."

The Ramalina group derives its name from a word signifying *dead branch*, probably in allusion to the decaying branches on which the members of this family grow. These are, like the Evernia, shrubby lichens, covered with powdery warts, and cottony within. The Ash Ramalina is the most frequently found member of the family, its fronds hanging in branched clusters from every decaying limb of the old trees, here and elsewhere. We find it not only on trees but on old park paling and barn doors; and I even remember procuring handsome specimens in quantity from the wood-work of an old wind-mill, which had long faced wind and weather in an exposed situation on the Wiltshire downs. It is called the Ash Ramalina (R. fraxinea, *Plate XVI., fig.* 3) because it is supposed to prefer that tree, though it by no means confines its favours to it.

The Bundle Ramalina (R. fastigiata, *Plate XVI., fig.* 2) is scarcely less common than its brother of the ash. It grows on trees and shrubs, bearing borderless apothecia in abundance on its short upright clumsy branches. We have gathered it in most of the counties we have any of us visited.

The Ivory lichen (R. scropulorum, *Plate XVI., fig.* 4) affects maritime rocks. We gathered some specimens of it on the Cornish shore, but found it in much greater beauty and abundance, draping rocks overlooking the sea, at Oban. Its apothecia are brownish and wart like. Its pale ivory hue makes it very attractive in appearance.

Succeeding the family of Ramalina comes the equally attractive one of Usnea, the members of which ornament decaying branches of trees with great beauty and profusion. Here the fronds are rounded, branched, and drooping, a central thread runs through them.

The Flower-like Usnea (U. florida) is perhaps the prettiest of the group. Its main branches grow as thick as fine cord, and have one or two lateral branches, all the branches are beset with numerous branchlets, crowding one upon another on either side the stem, varying in length from a few lines to half an inch. The lateral branches bear very large greenish ochre apothecæ, fringed like the branches with tiny branchlets; the branches grow in thick tufts, and they and their branchlets are covered with grey powder. This lichen is frequently found. We have specimens from these Trelawny woods, Cornwall, from York, Kent, and Wilts; but never did I see it in such abundance as in the Chace wood, near Ross, in Herefordshire. There the woodland paths were strewed with its fully branched tufts after each oft recurring gale. Old Gerarde writes of this "flouring branched moss," and thus describes it, " There is oftentime found upon old Okes, Beeches, and suchlike overgrown trees, a kinde of Mosse having many slender branches, which divide themselves into other lesser branches, whereon are placed confusedly many very small threads, like haires, of a greenish ash colour. Upon the ends of the tender branches there cometh forth a floure, in shape like unto a little buckle, or hollow mushroom, of a whitish colour, tending to yellownes, and garnished with the like leaves of those upon the lower branches."

The Hairy Usnea (U. hirta, *Plate XVI., fig.* 16) grows on dead branches, shrubs, and pales. It forms a sturdy little shrub, its branchlets thorn-shaped, and its whole aspect firm and strong. We find it frequently in the same habitats as its brother.

The Usnea Barbata, or Hairy Usnea, is found in old woods in hilly countries. It is jointed, very slender and long, and hangs like bunches of hair from the trees.

We found it in a glen on the Yorkshire moors, near Dallaghgill: the country people called it "young man's hair," because of its brown colour; while they designated the flower-like Usnea as "Old man's hair" from its white tint. It is said that this lichen is used as bandages for the feet in Lapland, and forms food for sheep and cows in Canada. The stringy Usnea (U. plicata) is found on park palings; it grows long and slender, and is of a grey powdery hue. I have it from Fowler's Park, Hawkhurst. Captain Weddell, in his journal, describes a species of Usnea in the New South Shetland Islands, "very beautiful, and bearing large deep chestnut fructification;" he speaks of "a little straggling grass, and a few mosses and lichens, as the only produce of these far southern lands."

The thread-like fronds of the Rock hair (Alectoria jubata, *Plate XVI.*, *fig.* 5) are to be found upon trees, firs in general; it only flourishes in alpine situations. It grows in mountainous districts in Britain; but the only time I ever found it was in Switzerland. Trees by the side of a snow torrent many hundred feet above the level of the sea, had all their lower branches bearded with long dark waving hair. These lichen fronds are branched and of a dark brown; the apothecia sessile and black.

The Cornicularia group succeeds that of Alectoria. Here the branches are shorter, very firm and shrub-like, and the shield-like apothecia are borne on the summits of the stems or branches.

On a wide moorland, forming the flattened summit of an extensive cliff, near to the small town of Polperro, in Cornwall, we found an elastic carpet of black matted threads beneath some overhanging rocks ; these threads were branch-like, and forked, and bore here and there little black knobs, which were the characteristic apothecæ of the family. This was the Woolly Cornicularia (C. lanata, *Plate XVI., fig.* 8). There are several species of this genus to be found in Britain : one, the prickly species, produces the crimson pigment which we call lake. The dark, the black and grey, and the sulphur species are all inhabitants of Alpine moors in Scotland.

The family of the Coral lichens comes next in order: to the naked eye they appear as Crustaceous lichens, but the microscope reveals the title to be classed among the branched group, and their receptacles are thus shown to be cup-shaped. The tree Coral lichen (Isidium heteromalla) grows in old trunks in the southern and eastern counties, and the granulated in similar situations. The speckled, the dotted, the white, and the eye-like species, frequent rocks in Scotland.

The Sphærophon group bears globular apotheciæ, and has a solid, stem-like, branched frond. Forests of the Coral-like species (S. coralloides, *Plate XVI., fig.* 9) adorn those Swaledale rocks, where the Cudbear and Crab's-eye flourish so luxuriantly. The stem is brownish and rather flattened, but the branches are grey, and are forked at the tips. It was brittle when dry, and its reddish brown apotheciæ were globular. The Compressed species (S. compressum) we gathered on rocks on the heights overlooking Loch Lomond. The crowded plants were very white and hoary, exceedingly minute, and looked as if covered with hoar frost.

The Stereocaulon family very much resemble the preceding one, but here the apotheciæ are flat. The stem of the Common Stereocaulon is rough, shrub-like, and beset with lateral branches, and crowded hoary foliage. It grew upon the rocks on the hills about Oban in great luxuriance, attaining two inches in height, and bearing numerous blackish apotheciæ (S. paschale, *Plate XVI., fig. 9*). It is a clumsy little plant, with thick stem and crowded branches; these are not hollow, but solid, hence its name. It abounds on mountains, and its receptacles are flat and sessile. It grows in high latitudes, along with the Rein-deer lichen, and the animals feed on it if their favourite lichen fails. There are several other species of solid-lichen, all more minute than the branched one, and chiefly inhabitants of Scotland.

And here on these sea-washed cliffs, as well as at the Cheesewring, and Starvegoose, and on the Yorkshire moors, grew the Rein-deer lichen (Cladonia rangiferina, *fig.* 11). It forms a white undergrowth beneath the heath and ling, or it covers dry plots with its interlaced branches. A pretty miniature tree, with round sessile borderless receptacles, it can endure any amount of heat and cold; even surviving the fires which so often burn up the heath. The Laplanders give thanks to God for this lichen, according to Linnæus. " A bounteous Providence sends us bread out of the very stones," they say. It is truly a great gift to them, for it supports their deer, and is eaten, when cooked, by themselves. What a lesson of gratitude to us! Do we thus heartily acknowledge the profuse gifts for convenience, and even luxury, with which our hands are filled, and for which this Lapland moss is but a very wretched substitute? Surely our teeming harvests and prolific flocks should rouse us to thanks, fervent and grateful as those of the Laplanders, and proportioned to our greater blessings!

The Rein-deer moss makes a matted carpet beneath the ling on moors. Its branches interlace, and often grow to the height of eight or ten inches, where the heath has been long undisturbed. Crabbe calls it most aptly:—

"The wiry moss that whitens all the hill."

Another species of Cladonia is common enough upon our heaths and moors, the Forked lichen (C. furcata, *Plate XVI., fig.* 10). It has a brownish, instead of a greenish hue, and its branches are simple, or only once forked. It grew at Starvegoose, on the heights near Palperro, and on the hills of Oban.

We now come to the charming group of Cup lichens, or, as they are more generally called, Cup mosses. These lichens have scale-like fronds, growing close to the stone or earth; a stem rises from these which bears a cup, the stems being called *podetta*. The cup in its turn supports the apothecia, generally around its margin.

The Common Cup moss (Scyphophorus pyxidatus, *Plate XVI., fig.* 10) is familiar to all who observe natural objects at all. I remember the delight evinced by a town-bred lady to whom I introduced the plant in Penyard Wood, Herefordshire. She had already filled her hands with flowering Broom, and Bitter-vetch, and Wood Anemones, but she poised the cluster of Cup moss on the top, to be treasured most, because it would not fade.

This is what old Gerarde speaks of as the "cup mosse," and the qualities of which he sets forth :—"The Muscus pyxidatus I have Englished cup mosse, or chalice mosse. It groweth on the most barren, dry, and gravelly ditch banks, creeping flat upon the ground like unto liverwort, but of a yellowish-white colour, among which leaves start up here and there certaine little things fashioned like a little cup, called a beaker or chalice, and of the same colour and substance as the lower leaves, which undoubtedly may be taken for the flowers; the powder of which, given to children in any liquor for certaine daies together, is a most certaine remedy against that perilous malady called the chin-cough." This lichen has been used as a remedy for hooping-cough long since the days of Gerarde. It is of a glaucous colour, and bears its fruit on tiny footstalks rising from the rim of the cup or chalice.

The Fringed cup-lichen (S. fimbriatus) is even prettier, for its chalice is delicately fringed, its colour more glaucous, and the tiny fronds of its thallus are lined with silvery white—these cluster not only on the ground, but on the podetia. The Thread-shaped cup-lichen has the smallest cups of all, scarcely more than the mere hollow end of the stem; the small apotheciæ are situated on the rim of these, and often quite fill up the cavity. All these we have found upon our Yorkshire moors, and near Callander and Oban, and among the Pentland hills. Near Oban, particularly, they flourished in great luxuriance.

But the minute tubercles of the thread-shaped species were quite thrown into the shade by the equally brilliant and much larger receptacles of the cochineal and finger-cup lichens (S. coccinea and digitata, Plate XVI., figs. 12 and 13). These were old friends, having been among the denizens of Starvegoose, with which we became acquainted on our first excursion. One of them is the plant described by Mrs. Hemans as suggesting so touching an association :—

> "And one, with cup all crimson dyed,
> Spoke of a Saviour crucified."

The horned cup-lichen closely resembles the thread-shaped species, but the tubercles were much larger in proportion to the cups (S. cornutus). None of the cup-lichens are prettier than the torn-coated one (S. sparassis, Plate XVI., fig. 11), with its clusters of leaves, or scales, growing on its cup or stem, and giving it such a finished appearance. This plant I found first in Kent, and recently in Scotland. This family is an especial favourite with the poets, Wordsworth launches forth in their praises when he writes—

> "All lovely colours there you see,
> All colours that were ever seen;
> And mossy net-work too is there,
> As if by hand of lady fair
> The work had woven been;
> And cups, the darlings of the eye,

> So deep in their vermillion dye.
> Ah me, what lovely tints are these,
> Of olive, green, and scarlet bright!
> In spikes, in branches, and in stars,
> Green, red, and partly white?"

But freer, fuller, handsomer than any yet described was the Daisy-like cup lichen (S. bellidoflorus) which we found near Ballahulish. A forest of podetia arose like flower stems, feathered from the foot with scale-like fronds, the cups were surrounded with double and treble radiating lines of foot-stalks, upon each of which a pink receptacle was situated; the heads reminded me more of hen and chicken daisies than of anything else. And the morning sun shining on the dew-bespangled lichen, and on the grey rocks around it, and the blue waters of the Loch, added tenfold to its charm, and impressed us with an idea that no lichen beauty can equal that of our daisy friend of Ballahulish. Can Mrs. Hemans have been there too!

> "Oh green is the turf where my brothers play,
> Through the long bright hours of the summer day;
> They find the red-cup moss where they climb,
> And they chase the bee o'er the scented thyme."

The Elegant cup moss (S. gracilis) grew by the Junction Road near Hawkhurst, and on Starvegoose also. It is a tall slender species, brownish in hue, and the podetia sometimes forked.

The Pycnothelia group has only one British representative, and it is so minute as to be little more than a crustaceous lichen. A greenish grey crust studded with pimples, of the same colour, tipped with brown, is all the charm it boasts. It is found in Norfolk, but we have been able to procure no specimen. Thus, then, we complete our present stock of lichens. We have collected diligently, and though we have failed in procuring specimens of some families, yet we have succeeded in far the greater proportion. The seed of these plants seems to be their most important and reliable feature, which seeds are contained in sacs, and the sacs lodged in a receptacle. Some sacs contain but one seed, some two, and some many (*fig.* A.) The receptacles are of various shapes and forms: goblet-shaped, as in the goblet-lichens: linear, as in the writing-lichens; wart-like, as in the wart-lichens, etc.; shields, as in the Lecidea and Lecanora, etc.; powdery warts, as in the branch-lichens; and tubercles, as in the cup-lichens. The form of the frond, too, is a good mark of distinction: crustaceous in the mushroom, goblet, writing, wart, leprous, internal fruited, and shield-lichens; the scaly lichens, with fronds powdery within, and leafy towards the edge, form a connecting link between the crustaceous and frondose parties; and the parmelias, lateral-fruited, dotted, gelatinous, socket, circular, and buckler-lichens are decidedly leafy in their habit. The last-named verge towards the branched lichens, and the ramilinas, usneas, hair, horned, coral, globe, and solid belong to the leafy order. The cup form is restricted to the one family of cup-lichens.

The Rev. W. A. Leighton has worked out another arrangement of lichens, dividing them by different characteristics. He calls the first tribe *Idiothalami*, such as have fruit clusters or *apothecia*, differing in colour and substance from the rest of the plant: to this tribe belong the Lecideas, Lecanoras, Gyrophoras, Endocarpons, and some others; the second tribe he calls *Cænothalami*, such as have the apothecia *partly* formed of substance of the *thallus* or leafy part of the lichen: to this division belong the Peltideas, Borreras, and other of the broad-fronded families; the third tribe he calls *Homothalami*, such as have the apothecia entirely of the same form and colour as the frond, and to this tribe belong the Usneas, Alectorias, Ramalinias, etc.; and he calls the fourth tribe *Athalami*, whose apothecia are absent and their fructification imperfectly understood, ranking under this head the Opographæ.

The rocks about Granton, and those on the shores of Arran and Ardrossan, furnished us with two small plants which we feel puzzled whether to assign to the sea-weed or the lichen group. Growing like a coating upon the stone, their flattened branches closely interlacing, blackened and crisp in the sunshine, they resemble minute fuci in form and colour; but their fruit was contained in orbicular tubercles, which seemed rather to claim relationship with the apothecia of lichens. We could find no gonidia grains, otherwise we should at once have proclaimed the two minute rock dwellers to be true lichens. These plants are called Lichinia, the larger species, small as a Screw-moss, grows between tide marks, and every flow turns its blackened fronds to a full olive colour (L. pygmæa). The smaller species grows high and dry; we found it crisp and brittle on the Cornish cliffs where it seemed as careful as the Samphire to avoid the reach of the high tides (L. confinis). Until we have authority for locating these little plants among true lichens, we will consider them lichen allies in the same way as we associate the Lycopods with the Ferns, and the Lungworts with the Mosses.

Plate 17.

1 Fly Agaric. 2 Purple A. 3 Hollow-stemmed Collybia. 4 Oak-leaf C. 5 Rosy Mycena.
7 Cup Omphalia. 8 Dingy Pleurotus. 9 Golden Pholiota. 10 Variable Crepidotus.
11 Verdigris Psalliota. 12 Olive-gilled Hypholoma. 13 Inky Coprinus. 14 Starry C.
15 Red Russula. 16 Orange Omphalia. 17 Pale Tricholoma. 18 Edible Mushroom.
19 Cornucopia Craterellus.

RAMBLES
IN
Search of Flowerless Plants.

CHAPTER XXV.

FUNGI.

"There's a thing that grows by the fainting flower,
And springs in the shade of the lady's bower;
The lily shrinks, and the rose turns pale
When they feel its breath in the summer gale,
And the tulip curls its leaves in pride,
And the blue-eyed violet turns aside,
But the lily may flaunt, and the tulip stare,
For what does the honest toadstool care!"
OLIVER WENDELL HOLMES.

THESE are no longer days of blindness. If the people in this world may still be divided into the two classes "eyes" and "no eyes," by far the greater number of the civilized community would be found in the first class. People's eyes are very wide open indeed, and there is an evident desire to see as much, and as far as possible. On our recent excursion to the Highlands we saw this eagerness constantly evinced. Coach passengers, steam-boat passengers, pedestrians, all were in a fever for *seeing*. Some were so thoroughly carried away by the profusion of beauty around them, that they forgot that they were seeing ; but the generality were bending every energy to follow in the steps of other men, seeing what had been described, and doing what others had done. Ladies recognized the Oak and Beech-ferns from the top of the coach ; others exclaimed upon the shamrock, the grass of Parnassus, and the Devil's-bit Scabious empurpling the meadows ; sprigs of the Bog-myrtle were gathered and passed from hand to hand, and many a weather-stained hat and bonnet was garnished with the Ling or the belated garlands of the Woodbine. Some even noticed the broad patches of lungs of the oak which half covered the bole of many a tree in the Trossachs, and other passes, and the copper coloured circles on the rocks in Lord Breadalbane deer-park attracted attention from a few. Birds, beasts, and in-sects ; rocks, streams, and mountains ; highland cottages, and the bare-legged children swarming from them, all received their full meed of observation. The beautiful, the curious, the useful, all were noted—all, but with one exception.

There is one group of flowerless plants to which our race is notoriously unjust. Show to a man of taste and poetry the tall column and spreading capital of the Fly Agaric, its cap covered with glossy crimson, flecked with torn fragments of its white felt swaddling clothes, " It's only a nasty toadstool," he says. Men will acknowledge beauty in the tiniest moss, the most formless lichen, or even in coarse sea-wreck, and then peep into your basket of Fungi, varied in form, and of every brilliant hue, and merely exclaim with disgust, "What a lot of toadstools!" Fungi are only accounted fit to be *kicked* over, hands are considered too good for them. "Is it a mushroom?" asks a boy, whose gastronomic justice is fully aroused, but when you answer "No," he forthwith upsets the "horrid toadstool."

Surely the time has come for the poor Fungi to get a fair hearing. As if knowing our tastes for strong contrasts in colour, and variety in form, they combine themselves in groups of endless variation in tint and contour. Out of the grass on the banks of those highland roads sprang black tongues, contrasting strangely with the soft surrounding moss, the passers-by praised the moss, but left the weird tongues unnoticed. It was the same with the yellow and violet coral-like branches clustering under shelter of the ling—these belonged to the fungi, so no one had eyes for them. Standing in their accustomed mushroom form, covered with scarlet, brown, or violet, or still more often orange, the tourists could praise trees, and rocks, and flowers, and even the mosses and the fallen leaves, but not one word for the gorgeous congregation of fungi!

Yet the fungi are marvellous in structure, attractive to the instructed eye from their grace of form and brilliancy of colour, but still more from the hidden marvels of their formation revealed by the microscope. They come, along with the mosses, when the fields and woods have lost all other attraction, and they give interest to the late autumn and early spring ramble, not withdrawing their graceful presence even during winter, if the frost spare their delicate structure.

Fungi, like other plants, have three parts, and these,

like the mosses and liverworts, are formed of cells. The *spawn* of fungi answers to *roots* in other plants; the growing part, be it umbrella-shaped, as in the mushroom, branched as in the clavaria, or cupped as in the peziza, stands for their stem and leaves; and the *spores* answer to their fruit. The cells are of various shapes, sometimes round or oval; sometimes, as in moulds, branched in a fanciful manner, so as to look under the microscope like crystal trees. They have two kinds of fruit; one where a seed or *spore* is formed at the end of the cells; the other where seed-like bodies, or *spordiæ* are formed *within* the cell. This distinction divides the fungus family into two divisions: 1st, the *spore-bearing* class, *Sporiferi*; 2nd, the *Ascus-bearing* class, *Sporidiferi*.

The first group of the Spore class has its *hymenium*, or fruit-bearing part, open to the air. The numerous members of the large group of surface-fruited fungi (Hymenomycetes) is ushered in by the enormous family of Agarics, the commonly accepted form of fungus or "toadstool." Here we have a columnar *stem* (stipes), hollow or solid. A *cap* like the top of an umbrella (pileus); *folds* (lamellæ) underneath the cap over which the cells constituting the hymenium are spread, upon each of which one or more *spores* are situated. Many of the Agarics are furnished with a *veil* (vellum), extending from the margin of the

1. Cell and Spore of Agaric.
2. Ascus of Peziza.
3. Spore of do. magnififd.
4. Hymenium of Agaric.

cap to the neck of the stem, and when loosened from the margin to allow of the dispersion of the spores, falling like a frill round the column; in some species, there is also a *wrapper* (volva) which envelopes the fungus in its infancy. This Agaric family contains upwards of a thousand species, scientifically distinguished by the colour of the spores.

This sounds an easy mode of distinction, and we gather the fungi, and search for the seed; but at first we find none; then flying to a new conviction, we proclaim that there is no seed, so we cannot distinguish its colour. Let us take a ripe Agaric, cut the stem off on a level with the margin of the cap, and place the head on a sheet of paper. Lift it after some hours, and behold a perfect delineation of every fold in fine powder, deposited star-like on the paper. This powder consists of millions of spores; they may be white, yellow, brown, or purple; subjected to the microscope, they are all shown to be symmetrical in form, round in some species, oval in others, but all alike. Nor is the Agaric exhausted: subject a section of one fold to the microscope, and you see it covered with pimpling cells, and quartettes of spores adhering to the cells over great part of the surface. If the first crop of seed is shed, the second is ripening; and when the head is removed, it will proceed to deposit a star of dust.

The first time when fungi became really objects of interest to us was one beautiful autumn day when a party of us were spending a day at that pretty little lake on Lord Bath's property called Sheerwater. Not caring to go in the boats with the majority of the party which composed the pic-nic, we set off to seek for plants, flowers if possible—and if not, ferns or mosses. A few lingering sprays of the Lesser Scull-cap rewarded our search, but no new ferns, although the Spreading Shield-fern was flourishing in such extraordinary size and elegance, that we were half inclined to erect it into a new species. We collected a few mosses, but these were of the kinds most frequently found; and were on the point of grieving that the botanical season was over, when we came in sight of a number of Fly Agarics (Amanitus muscarius, *Plate XVII., fig.* 1) dotting the ground beneath some beech trees, and exhibiting every stage of growth, from chestnut-sized buttons, closely enveloped in their felt-wrappers, to umbrella-shaped, half-grown specimens, the thick veil just breaking from the edges of the cap, and the crimson surface closely dotted with the torn fragments of the wrapper; and full-grown stately tables, the glossy crimson skin almost or quite free from the scraps of white felt, and the veil gone, except for a frill round the neck of the stem. We exclaimed at the beauty of the plant, regretting that they were not such as we were in search of, and would therefore afford no pleasing occupation for our pencils on the morrow, nor any memorial of our happy day, until the wisest of our party suggested that we should begin to draw the fungi too; and thus a fresh subject of interest was brought before us, which has been a source of great happiness to each who that day entered upon the pursuit of it.

How many scenes does that gorgeous Agaric recall to my memory. A gentle lady, driving through the stately plantations surrounding the country-seat of Mr. Beresford Hope—"Oh, how beautiful," I exclaim; and she, with ready kindness, calls to her coachman to stop, and let me alight, and collect my treasures. I did not want the fungi, though I could not restrain an exclamation at their beauty; but now I gather them to bring to her. She is charmed and amazed to behold anything so beautiful in toadstools! Again, the secluded grounds of a true lover of nature rise before me, and the Fly Agaric grows freely beneath the spreading beech trees. The owner, though accounted a matter-of-fact lawyer, goes every morning to look upon the brilliant fungi, which have put themselves under his protection. A professed devotee of nature and poetry visits the place, and the master shows his lovely fungi with as much pride as his elm avenue and yew grove. He might with truth say with Christina Rossetti :—

> "All caterpillars throve beneath my rule,
> With snails and slugs in corners out of sight;
> I never marred the sudden curious stool,
>   That perfects in a night."

But the eye of the stranger is blinded by conventionalities, and he passes the Agarics afterwards with a sneer, "Be careful not to upset Mr. ——'s pet toad-stools."

This plant is poisonous in England, and is used for exterminating flies and vermin; in other European countries it is merely intoxicating, not poisonous, and is used as an article of food in Kamschatka. This alteration of quality with climate is not uncommon among fungi. Many species which are deleterious here are regular articles of food upon the continent, and the Fuegians depend greatly upon some for sustenance which are actually poisonous in Britain (Amanita muscarius, *Plate XVII., fig.* 1). The Red-fleshed Agaric (Amanita rubescens) is the near relation of the Fly Agaric. Its cap is brown, and its flesh turns red when bruised. The wrapper adheres in this species just as in the one which we have been describing; and it is as frequent, though less showy, an ornament of our woods. Berkeley describes its qualities as "doubtful;" but I have heard it recommended as edible. I have found it in Yorkshire, and frequently in Kent.

A third Amanita (A. phalloides), the Phallus-like Agaric, greeted our sight when spending a day at Virginia Waters. London was crowded and oppressive in the extreme; we had toiled at the exhibition of '62 day after day, and when our humane host proposed one day in the country, to recruit our exhausted powers, we eagerly entered into his plan. By that calm lake we sat and luxuriated, gathering several varieties of heath, and the Lesser Dodder parasitic upon it, treading over thick mats of the White-thread moss, and filling the provision-basket, which we had quickly emptied, with fungi of various shape and character. One of our group was this Phallus-like Agaric. The wrapper did not adhere to the cap, but opened at the summit, and let the tall plant shoot from it, remaining like a soft egg-shell round the base of the stem. The cap was yellowish green, very glossy, the veil thick and entire. There are other species of Amanita, but they have not rewarded our search.

Another group of Agarics where the veil is very fully developed is characterised as *Lepiota*. The Bulbous Lepiota grows in rare beauty under a fir avenue in a lovely pasture near Hawkhurst, in Kent. We have found them there year by year. The stems tall and bulbous at the base, the pale caps umbrella-shaped in youth, but tabular in maturity, and all beset by brown scales arranged in circles, from the centre, formed of remains of the outer skin. This Agaric is doubtfully wholesome, though its brother, the A. procerus, which it closely resembles, makes fair ketchup, and is sold for that purpose in Covent Garden Market.

The group of Agarics designated as Tricholoma have the folds sloped off to the stem, and the veil is woolly, and adheres to the margin of the cap, instead of forming a ruffle round the stem; in very many of the species it is absent altogether. The White Agaric (T. albus), which so often attracts our attention, in woods, by the snowy or buff tinted colour of its ample cap, belongs to this group. This was in its glory among the sward at Virginia Waters. It flourishes luxuriantly in the Bedgebury woods, and is found equally in the north of England and Scotland.

One of the fairy ring Agarics (T. personatus), grey in colour, and with a bulbous stem, is also sold in Covent Garden. The stem is tinged with lilac, and is short, and the plant has a solid appearance. We have met with it on the Downs and in pastures.

The large brown T. grammopodius is a frequent and showy ornament of wooded pastures. It grows to a large size, and the centre of the cap is raised and tinted with a deeper amber than the circumference. The Kentish wood borders are often beset with it.

The delicate mouse tint of the Pale Tricholoma (T.

humilis, *Plate XVII., fig.* 17), first greeted us among furze and heath upon the Wilts Downs. It is a tall, stately fungi, three or four inches across, and attractive from its elegance of form.

The same woods furnish abundance of a small Agaric, belonging to the next, or Clitocybe group, which is attractive from its fragrant odour (C. odorus). It grows among moss, is of a pale greenish mouse colour, and measures about two inches across. The Bell-shaped Agaric (C. Cyathiformis)—of similar size, but slenderer growth, its cap depressed into a funnel shape—rewarded our search in the woods and field borders about Callander. It is a curious species, from the uncommon form of the cap.

The little Purple Agaric (Colybia laccatus) is a frequent ornament of our woods, its rich tint contrasting beautifully with the fresh green of the surrounding moss (*Plate XVII., fig.* 2). Its colour varies much, though always rich—now amethyst purple, now maroon. We first found it on that memorable excursion to Sheerwater, which occasioned our beginning the study of fungi.

In the Collybia group the margin of the cap is at first rolled in, but this feature often disappears early in the life of the plant. The Hollow-stemmed Collybia (C. fusipes, *Plate XVII., fig.* 3), I first saw growing in a cluster from the bole of a tree, in the Hope Park, Edinburgh. Of course I was all anxiety to procure the specimen, but the ground where these trees grew was fenced off with park paling, and I saw no chance of reaching the fungi with my own hands. Presently, however, a "laddie" appeared, and readily undertook to scale the fence and seize the objects of my desire; though what reason I could have for wanting "puddock stools" it was utterly beyond his power to conceive. This species has an inflated stem, often twisted and cracked; the whole plant is of a reddish brown in maturity, though, in an earlier stage, the folds are of a pale tint. The Velvet-stemmed Collybia is a very handsome fungus. We have gathered it from the stumps of trees in autumn in Wiltshire, Kent, and Berkshire. The cap is a rich burnt senna colour, the folds yellowish, and the stem senna, shading to black, and covered with a velvet pile; it has rootlets at the base. An edible species—with mouse coloured cap, and stem, and pale folds, which we found in the Braid glen near Edinburgh—is the Nail Mushroom, which Berkeley describes as an article of commerce in Austria (C. Esculentus). The Oak-leaf Agaric (C. dryophilus, *Plate XVII.*, *fig.* 4), is a pretty species, and often found among decaying oak and beech leaves in the Kentish woods. The whole plant is of a buff colour.

In the Mycena group the margin of the cap is straight, and pressed to the stem in infancy. The pretty rose Agaric (M. Purus, *Plate XVII.*, *fig.* 5), is frequently found in woods especially under larch and fir. I remember coming upon a perfect crowd of it in the plantations at Starvegoose, near Hawkhurst. The folds are very broad, the margin of the cap striped at the edge, and coloured more deeply at the centre; and the stem has root-like hairs at the base. The whole plant is a beautiful lilac colour. A minute and slender Agaric (M. alcalinus) familiar to all observers of nature, belongs to this group. It appears in a night, raising its slender bells by dozens on decaying wood, while the tall stems that support them are so slender that they shake and quiver in every breeze. The colour is grey, shaded dark to the summit of the bell, and pale to the margin; the folds are white at first and then become greyish. The smell is like that of walnuts; and the plants perish as rapidly as they appear. Another pretty species of this group, sent to us from the Somersetshire woods, is the pale green Mycena (M. epipterygius), it grows among fern leaves, is decidedly green in hue, and in other respects resembles its allies.

On a twig of bramble I found a cluster of slender Agarics, their stems thread-like, their caps pure white. The twig was in a decaying state, lying amongst fallen leaves in the Chase Wood, near Ross, in Herefordshire. The fungus was the bramble Mycena (M. roridus).

The folds turned down the stem mark the Omphalia group. A pretty species of Omphalia attracted my attention as it grew from the margin of a peat-bog on the Yorkshire moors, very early in my fungus experience. It is a small plant, not measuring more than two inches across; the cap depressed in the centre, so as to earn for it the name of Cup Omphalia (O. Pyxidatus, *Plate XVII.*, *fig.* 7), and striped and puckered at the margin. The colour of the whole plant is a bright ochre.

The Orange Omphalia (O. fibula, *Plate XVII.*, *fig.* 16) is a slender and elegant species. We found it in the grounds of Craig House, near Edinburgh—parasitic on a dead leaf.

The Pleurotus group has either a one-sided stem or no stem at all. There are some very handsome species, as the Meadow and Oyster Pleurotus (P. ulmarius and ostreatus) which grow bracket-like from the boles of trees; but none of these have rewarded our search. Only the Dingy Pleurotus (P. applicatus, *Plate XVII., fig.* 8) graces our collection. It is very minute, dark, dingy purple, and stemless. All these groups of Agarics have *white* spores.

We now come to a series with salmon-coloured spores, the first group of which, Volvaria, is characterised by a distinct wrapper as in our old friends, the Amanitæ. Of this group we have been able to procure no specimen.

The Pretty Plum Agaric (Clitopilus prunulus), of pure white colour and esculent qualities, belongs to one of the early groups in this series. The folds are turned down the stem; and though white at first, like the rest of the plant, they become flesh coloured as soon as the spores are ripe. We have found it in woods in Wilts and Kent.

No Agaric is more attractive in its season than the Golden Pholiota. The group to which it belongs is the first of the brownish spore series.

This fungus (P. aureus, *Plate XVII., fig.*, 9) is frequently found about decaying stumps, its brilliant yellow contrasting advantageously with the dead hue of the wood, or the green mosses adhering to it. It is one of the most frequent of our fungi.

The Hebaloma group are distinguished by a scaly stem, but we have not been able to find any of its representatives.

The Naucoria group has an hemispherical cap. One little brown species (N. melinoides) is very common on lawns, its head no larger than the end of a child's finger. Another, the Branny Naucoria (N. furfuraceus) frequents thatch, growing alone or in groups, the pale brown cap covered with silky scales.

The Crepidotus group comprises a number of small fungi, many of them less than an inch across, with lateral stems, or no stems at all. The Variable Crepidotus (C. variabilis, *Plate XVII., fig.* 10), is very common on dead sticks, and we have found it in every county where we have searched for fungi.

We now come to a series where purple mingles with the brownish hue of the spores. The first group of these Psalliota, contains our much respected edible mushroom, (P. pratensis, *Plate XVII., fig.* 18).

> "Then sleep the seasons, full of might,
> While slowly swells the pod,
> And rounds the peach, and in the night,
> The mushroom burst the sod."

How shall we sufficiently praise this popular plant! Whether gathered in its infancy as "buttons" for a pickle, or in its mid age, for frying or stewing, it is a delicious addition to the chop or cutlet, and we do not believe in its effects being harmful, if taken with other food. Berkeley gives this direction with reference to all edible fungi, to eat them with plenty of bread, or other simple food, most of the instances where they have had an evil effect have been when taken alone. Last autumn our daily rambles in Kent were professedly in search of mushrooms, and we seldom sat down to dinner without a dish of our gatherings, yet none of our party suffered in the slightest degree. Sheep are fond of mushrooms, and we found it seldom repaid us to seek them in fields where sheep were browsing. It is a curious phenomenon, the suddenness with which all fungi appear in a locality, often following upon some new dressing of the land. In a pamphlet on the botany of Gloucestershire, I saw a note illustrative of this. An intelligent farmer gave a heavy dressing of salt to a field of grass which had been frequently flooded by an adjacent brook, the inundations of which had rendered the grass sour. The following autumn the field was covered with mushrooms, "one person alone sold £20 worth of buttons for pickling." In this and analagous cases the probability is that the spawn is already present in the soil, but needs some extra principle to enable it to spring into vitality. In the Gloucestershire pasture, the wanting element was evidently supplied by the salt.

The Field Mushroom (P. arvensis) is a large and coarse species, also edible, and much sold in Covent Garden for stewing.

The Verdigris psalliota (P. æruginosus, *Plate XVII.*, *fig.* 11), is remarkable for its rare tint. The full glaucous green is very seldom found among fungi, and we hailed it with triumph one autumn morning when we were gathering hops in a Kentish lane. In youth this fungus is conical, but spreads as it nears maturity, but the raised centre of the cap never becomes plane. The hemispherical Psalliota (P. semiglobatus) is a frequent denizen of stubble fields and pastures, its cap brown, and its folds filling in the cupola formed by the rounded head—the spores here are darker than in the other members of the group.

The Hypholoma group is characterised by a web-like

veil, which disappears altogether before the plant attains maturity.

The olive-gilled Hypholoma (H. sublateritius, *Plate XVII., fig.* 12) is a handsome species, the cap bright red brown and convex, the folds pale at first, then olive. Our specimens came from the Crockerton woods in Wilts. The Bundle Hypholoma (H. fasicularis) is a familiar example of the group. Clusters of it grow out of dead wood, hazel, and other branches, at most seasons of the year. The colour is orange, and the folds are only varied by a slight admixture of green.

The next series of Agaricini have black spores. The Coprinus group spring quickly, and then melt away. The Hairy Coprinus (C. comatus) has the cap torn into hair-like scales. It grows in a cone-shape, and is found on road sides and pastures. I first gathered it in a field at Kingston Deverill in Wilts, but I had scarcely time to paint its portrait before it melted away. It is a common species. The Inky Coprinus (C. atramentarius *Plate XVII., fig.* 13) is also conical in form, and grows in extensive groups, often crowding so close upon one another, as to push each other out of shape. It springs on garden borders, lawns, and in any rich ground. All the group live principally on manure.

About old stumps we frequently find similar groups of the Mica Coprinus (C. micaceus). It is so called from the sparkling scales upon the cap, which resemble grains of Mica, the cap is an ochre brown, and the stem is hollow. It is smaller and less conical than the other two species. The Folded Coprinus (C. plicatilis) is a common fungus on horse dung in pastures, very slender, the cap very delicate, and striped and soon splitting into ribbons.

But of all this group of Agarics, the Starry Coprinus is by far the most curious. The tiny fungus springs from radiating threads, and, in youth, resembles a tiny puff ball. It grows on plastered walls, the radiating spawn is necessary to collect nourishment for the plant from the atmosphere (C. radians, *Plate XVII., fig.* 14). My specimen grew on plaster only a twelvemonth old, and such was the interest of my gentle host in the study of fungi, that he did not grudge to chisel out the piece of plaster, to endow my collection with the specimen.

The Hygrophorus group has waxy folds, and the whole plant is of a brittle watery nature. The Meadow Hygrophorus (H. pratensis), is bright yellow or buff, moist and shining, and very ornamental. The conical species (H. conicus) is larger, of a more permanent yellow, shading to bright scarlet, very showy and attractive. The Paroquet Hygrophorus (H. psittacinus) is beautifully tinged with green both on the cap and folds, it studs the Wiltshire Downs with its gay cupolas, but its smell is offensive. In the Lactarius group the folds are milky, and the juice sharp and biting. The delicious Lactarias (L. deliciosus) is very handsome, apricot coloured, depressed in the centre, and of great size. Wherever it is wounded, a red juice oozes, but this juice is not acrid. This species is edible. I have found it in plantations about Hawkhurst.

The same woods also produce the magnificent Red Russula (R. rubra, *Plate XVII., fig.* 15), the most splendidly-coloured of the group. Here the cap is of intense carmine, and the folds and stem are white. The Emetic Russula (R. emetica) is scarcely less handsome, the colour being almost of the new magenta tint, varying in every shade from white to the deepest hue. The cap is expanded, not cupola-shaped, as in the last species; and the fungus is very unwholesome. It abounds in field borders, called in Kent shores, about Hawkhurst.

We now come to a group with branched and swollen folds, blunt at the edges, and rather like veins than agaric folds. The Chanterelle of the French, Cantharellus of Britain, has one edible species (*Plate XIX.*, *fig.* 12), Cibarius. This fungus is apricot coloured, variable in form, and very abundant where it grows at all. I first saw it studding the grass under trees at Virginia Waters; and I saw it, later in the same year, in tenfold abundance in the Bracklyn woods near Callander. The tawny Chanterelle (C. tubæformis) I found in Wiltshire; the veins are thicker, and more distant than in the edible species, and the stem is compressed. It is the scarcest species of the two. There is a pretty slender species, parasitic on moss or thatch, and elsewhere, the moss Chantarelle (C. muscigenus); but it has not rewarded our search.

A curious genus, with waxy veins, often growing parasitically on other fungi, is termed Nyctalis. The Starry Nyctalis I found on some dead agarics, near Hawkhurst; but the folds were not fully formed. Berkeley tells us, that the meal powdering its cap consists of starry bodies, seeming to be a second kind of fruit: hence the specific name.

The marasmius group contains the true fairy-ring

Agaric; an edible species, very abundant in some localities (M. oreades). In this group the substance of the folds is corky, and very dry; and the specimens are much longer lived than in most of the other families. For long the fairy-rings were an unexplained phenomenon; but now that fungus rings are so constantly seen on the Downs and elsewhere, in the vicinity of the fairy rings, it becomes self-evident that these plants are the agents in their formation. The manner in which they work is less apparent; but is probably that suggested by Berkeley and others. Throwing out the seed in quantities, so as to form a mass of felt-like spawn, the fruit probably springs at a certain distance from a common centre; and as the small circle will again cast seed beyond its own circumference, the ring naturally widens with each successive crop. The decayed substance of the fungi fertilize the ground, so that the fairy ring is constantly marked by luxuriant verdure.

But, as Miss Godewin says,

"The fairies long since trooped away,
Then fled the ghosts in full array,
And now each muse departs;
Expelled from grove and sacred stream,
Where erst they dwelt in airy dream,
The poor things break their hearts."

Here is a gloomy picture; and fancies should at any rate be cheerful. We turn with pleasure to Allan Cunningham's mention of the fairy ring.

"Oh, lead me forth o'er dales and meads,
E'en as a child the mother leads;

> Where twin nuts cluster thick, and springs
> The thistle with ten thousand stings;
> The ring where last the fairies danced,
> The place where dank Will latest glanced,
> The stream that steals its way along
> To glory, consecrate by song:
> And, while we saunter, let thy speech
> God's glory and his goodness preach."

A great number of minute and elegant parasites belong to this group.

From the decaying wood of a turnstile in Wiltshire, a group of almost stemless fungi were growing. The cap was of delicate brown, kidney-shaped, flat, and much curled and plaited at the edges; the folds were broad, in many instances torn, and of a pale colour. The substance was fleshy. This was the plaited Lentinus (L. flabelliformis), a rare species, of which our collection is proud in no small degree.

The little Styptic Panus is my only specimen of the Panus group; tough in texture, pale, mealy externally, its short stem swelled where it joins the head. It has styptic qualities, hence its name; and is found on dead trees and stumps, growing in shelf-like clusters. Our finest specimens are from Herefordshire.

The last group of Agaricini is the Lenzites, of corky texture, and folds branching and crossing one another. The Birch Lenzites (L. betulina) is common on old stumps and rails; it is hairy, and zoned with ochre and olive; the original tint being buff. It becomes almost woody in age, and lasts upon the tree stumps for a great length of time.

Old Gerarde takes strong views on the subject of fungi, and follows the rule which we may often see exemplified in the present day, that the farther from truth the opinions are, the more positively they are advanced. " Mushrums come up about the roots of trees in grassie places," says the old herbalist; "and by land newly turned in woods, where the soil is sandy, and yet dankish. They grow, likewise, out of woode, foorthe from the rotten bodies of trees ; but they are unprofitable, and nothing woorthe. 'Poisonous mushrums,' as Dioscorides saith, " grow where old rusty iron lieth, or cotton clouts, or near to serpent's dens, or rootes of trees that bring foorthe deadly fruit." Divers esteem those far the best that grow on mountains and hilly places ; as Horace saith :—

'The meadow mushrums are in kind the best ;
It is ill trusting any of the rest.'

' Mushrums,' saith Pliny, 'grow in showers of rain ; they come from the slime of trees.'"

CHAPTER XXVI.

FUNGI.

"The Turf
Smells fresh, and rich in odoriferous herbs
And fungous fruits of earth, regales the sense
With luxury of unexpected sweets."
COWPER.

WE pass from the extensive order of Agarics to one of much smaller dimensions, the second in the class characterised by the exposed hymenium. Here this fruit-bearing stratum is spread within *pores* instead of folds, and the order hence called *Polyporei*.

The Sap-balls are the first group in the Pore order. When disposing of fungi in the usual insulting manner of upsetting, we expect the reversed side to consist of folds in the regular toadstool pattern, but we are occasionally surprised (if surprise will condescend to be aroused by a fungus!) to see the interior filled as with sponge. The yellow Sap-ball (Boletus luteus, *Plate XVIII.*, *fig.* 1), is as bright in its golden hue as the orange Pholiota, and grows to a larger size. We have frequently found it about Hawkhurst late in the summer, and also in Herefordshire and Yorkshire. Its favourite habitat is fir woods. The White Sap-ball (B. laricinus)

Plate 18

1 Yellow Sap-ball 2 Lurid do 3 Moss Cyphel 4 Scaly Polypore 5 Fir P 6 Corky Merulius 7 Fistulina 8 Spread Hydnum 9 Purple Corticium 10 Brown Hymenochaete 11 Yellow Clavaria 12 Amethyst C. 13 Furrowed C. 14 Candles 15 G 16 Orange Tremella 17 Glandular Exidia 18 Jew's Ear 19 Bloody Stereum

is scarcely less common, it generally graced our fungus basket on our Kentish excursions.

Once only we found the Peppery Sap-ball (B. piperatus), the pores are honeycombed and much larger than in the other species, and of a pale yellow, while the cap is red brown. Our specimen grew in the Bedgebury woods. But the first Sap-ball which greeted our awakened observation was a tenant of the woods about Sheerwater in Wiltshire, and formed one of the first basket of fungi we gathered. It was an evil-looking species, well earning its name of "Lurid." The cap was umber and powdery, the tubes and swollen stem vermillion shading to black. It is very poisonous (B. luridus, *Plate XVIII.*, *fig.* 2). The Edible species (B. edulis), we also found in Wiltshire, but the specimens were bad. More recently we have found it in abundance in Kent. Here the cap is brown, the pores pale at first, and then olive, and the plant grows to a large size. The Scaly Sap-ball (B. scaber), is also a common inhabitant of fields and woods in Kent, it also is brownish with yellowish pores, and a very marked veil.

In the Sap-ball group the pores were separable from one another, in the succeeding, the Polyporus group, they are not separable. The Scaly Polypore (P. squamosus, (*Plate XVIII., fig.* 4), is familiar to all observers, growing like irregular brackets of great extent, and considerable bulk, on ash and other trees. We found our first specimen in the woods of Studley Royal, in Yorks, and presently afterwards saw it growing from trees in woods and hedge-rows in the more northerly districts of the same county. Also we have seen it abundantly in

Kent, and occasionally in Shropshire. The Shell Polypore (P. conchatus), I found on pollard willows on the banks of the Wye near Ross, in Herefordshire. It is an elegant species, partaking the character of the cornucopia in form, the cap deep red brown, and the pores varying from buff to brown. This species is much less solid than the last, owing to the shortness of the pores, and much more elegant in shape. The Common Willow Polypore was flourishing on the same trees, shapeless in form, pallid in hue, but the closely packed woody pores of a bright cinnamon colour. The P. annosus we found in the Sheerwater woods, adhering to large stumps, and growing quite into the ground. The pores were white, and the covering brownish. The variegated Polypore (P. versicolor), is a very familiar species; smaller than any of those already named, but growing in great abundance, tier above tier; zoned with olive and ochre, like the lenzites, and often half covered with a bright green tint, the result of the growth of minute confervæ. The Fir Polypore (P. abietinus, *Plate XVIII., fig.* 5), is smaller than the variegated one, but resembles it in character, it is of a beautiful violet hue underneath, white above, with violet edges. We have gathered this in Wiltshire and other counties.

The Trametes group is closely allied to that of the Polypores, only distinguished by the presence of *trama*, a stratum intervening between the fruit-bearing surfaces of folds or pores. The Corkey Trametes (T. suaveoleus), is of the texture indicated by its specific name, it is rather soft and pale, has large pores, and is powdery. We found one specimen in a hollow willow in Kent.

The Dædalia group have toothed pores. The Oak and One-coloured species are both common (D. quercina and unicolor), they are heavy shapeless fungi, growing shelf-like on trees and stumps and would be taken for Polypores, but for the curious form of their large pores. The name signifies a labyrinth, owing to the twisting habit of the pores.

The Merulius group is rendered notorious in fungus history for its evil member the Dry-rot (M. lachrymans.) The spawn of this fungus makes its way into the woodwork of houses, especially where the air is excluded, and soon reduces it to rottenness. We passed a church which was all gutted, fresh wood work in course of preparation, and in answer to our inquiries as to the cause of these restorations, we learned "that the Dry-rot had got into the wood, and it was all rotten." The same process was going on in the house of the Curator of the Edinburgh Botanic Gardens, and from the same cause. The substance of this fungus when fully developed is fleshy and watery, pale in colour, and the under surface velvety. Drops like tears often ooze from it, hence its specific name; it is only *dry* in its undeveloped stages.

In this group the hymenium forms netted folds, blunt, shallow, and vein-like. The Corky Merulius, (M. corium, *Plate XVIII., fig.* 6), grows commonly on stumps, is about the thickness of parchment, and but a shade darker in colour. In its early stage it is spread over the surface of the wood, but it presently becomes loose at the edge, and turns over in frills.

The last group in the Pore order is that of Fistulina. Here the pores are not in veins, or netted folds, but in deep tubes, quite distinct from one another. Our one British species (Fistulina hepatica, Plate *XVIII.*, *fig.* 7), is a scarce fungus, we have found it growing from trees in Longleat Park, Wiltshire; and some of our friends have it from Stoneleigh Park, Warwickshire. These good people were adventurous, and having heard that the unsightly fungus was as wholesome food as the beef-steak which it so closely resembles, they ordered it to be cooked, and proceeded to eat it. They described the flavour as similar to that of veal cutlet, and they experienced no evil effects from partaking of the dish. Their relatives were much agitated by the transaction, and the adventurers were awoke at daybreak, by inquiries regarding their state of health.

The small order of Hydnei succeeds that of Polyporei, its distinguishing feature being the *spines*, over which the fruit-bearing stratum is spread. The Spread Hydnum (H. repandum, Plate *XVIII.*, *fig.* 8) is edible. Berkeley describes it as delicate and agreeable in flavour, and quite wholesome. We have found it abundantly in the Bedgebury woods, and more sparingly in those of Wiltshire and Herefordshire. The spines thickly besetting the under surface of the cupola remind one of stalactites in a cavern. In Kent we found another pretty species, the zoned Hydnum (H. Zonatum). This is a much smaller

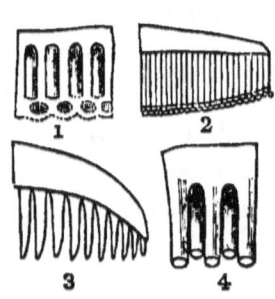

1. HYMENIUM OF POLYPORUS.
2. DO. OF BOLETUS.
3. DO. OF HYDNUM.
4. DO. OF FISTULINA.
All Magnified.

fungus, of a reddish like hue, marked on the surface of the cap with darker zones, the centre becoming depressed. The same woods, and those of other counties, furnished us with pale encrustations on dead sticks, which a pocket lens showed to be crowded with minute spines. This was the Ochre Hydnum (H. ochraceum) one of the stemless species. Several other groups belong to this order, but their members are few, and scarce, and none of them grace our collection.

The fourth order of the Hymenium class is called Auricularini, or Ear-fungi, and is characterised by the fruit-bearing surface being straight, or nearly so. In the first group, that of Craterellus, the Hymenium is straight in youth, but becomes rutted in age.

One splendid member of the group, the Cornucopia-like Craterellus, rewarded my search one autumn in the Chase Wood, Herefordshire. Half a foot long, and measuring five inches across the opening, the elegant form and sombre tints of the plants at once attracted my attention. They were growing in a group of three or four, near a tree root, and were everywhere shaded to black (*Plate XVII., fig.* 19).

In the Thelephora group the plants are scarcely more than encrustations, the surface sometimes rises into pimples. The ground Thelephora is brown and flattened, we found it in Wiltshire.

The Stereum group is also composed of mere thin corky encrustations. A lilac felt on poplar trees is one species (S. purpureum), a rich velvet pile spread on stumps, or turning over in frills and lappets, and shaded from orange to amber, is another species (S. hirsutum,) whilst little circular violet-brown patches, zoned externally, and turning crimson when wounded, present plants of the Bloody Stereum (S. sanguinolentum, *Plate XVIII., fig.* 19, right corner of the plate).

The Hymenochæte group is characterised by stiff bristles. The rusty species (H. rubiginosa, *Plate XVIII., fig.* 10) grows on posts and pales, is of a rich burnt siena colour, and looks like a folded piece of worsted velvet.

In the Corticium group the Hymenium is swelled when moist, and often fringed with tiny hairs around the edge. The Purple Corticium (C. cæruleum, *Plate XVIII., fig.* 9) looks like a morsel of rich blue velvet when moist, but it becomes duller when dry. The Oak Corticium (C. quercinum) is brownish lilac, and is very common on dead or aged oak branches, the patch measuring from one to three inches in length. The Elder Corticium, (C. sambuci) is white and very thin, it grows on Elder stumps, cracking with the inequalities of the bark. All these species are very common, the localities where we have found them are too numerous to mention.

A group of minute and elegant fungi form the Cyphella group, they are cup shaped, and often pendulous, somewhat resembling Cantherellæ in their veined Hymenium. Near Richmond in Yorkshire I found the little moss Cyphella (C. Muscigena, *Plate XVIII., fig.* 3) growing upon one of the large Feather mosses; this is the only species which we have any of us found.

We now come to the fifth order, that of Clavariæ, or Club-fungi, characterised by having the Hymenium upon the Club, up to its very summit. The true Clavariæ are fleshy and generally branched, the stem being the same in substance as the club. There is no prettier group in all the fungus kingdom than this. Its members vary in colour to every shade of purple and yellow, or pure white, and they grow either in single clubs, clusters of clubs, or coral-like branches, interlacing in every fanciful form. The Bundle Clavaria (C. fastigiata, *Plate XVIII.*, *fig.* 11) is extremely common on the chalk downs, and in pastures, it is often called the Coral Fungus, and the form of its branches entitles it to the appellation. It is repeatedly branched, the branches entangling with one another, and too brittle to be easily disengaged. The Crested Clavaria (C. cristata) is equally pretty in its way, very much curled and crumpled in its early stage, so as to resemble pieces of peeled walnut, but spreading into elegant snow-white branches as it approaches maturity. We have found it in woods in Wiltshire, Herefordshire, and Kent. Perhaps no species is more common than the Furrowed Clavaria (C. rugosa, *Plate XVIII., fig.* 13) it grows in single clubs, often dilating towards the summit, and bending into every quaint variety of position. Last autumn I gathered it in Kent of an unusual size, and in great quantity; some Shropshire friends gave a similar account of its abundance in their neighbourhood, and added, that they had had it fried like mushrooms, with butter and pepper, and that the dish was scarcely accounted inferior to the true mushrooms. The Violet Clavaria (C. amethystina, *Plate XVIII., fig.* 12) we found at Oban. It was a foggy day, and the grass and ling were heavy with moisture. But we felt that we might not be favoured with better weather during the week of our stay, especially as we had rashly used up two fine days in examining the treasures of the shore. So we were resolved to traverse the uplands, and endure the wetting as best we might. But it was heavy work! Our dresses were soon wet to the knees, and what a weight it was to carry. We toiled for hours, and were feeling quite discouraged, when my eye caught sight of a cluster of violet branches among the sward at my feet. Saturated as everything was with moisture, the colour of the fungus was in its full perfection, and lovelier shades of violet I never saw. A tawny Geoglossum was growing hard by, but of that more anon. The Candle Clavaria (C. vermiculata, *Plate XVIII., fig.* 14) is formed of a simple club, white, and generally congregated in clusters, which look like half a pound of miniature dip candles. It grows in pastures pretty frequently. I have oftener found it in Kent than elsewhere. Another species somewhat resembling this, but bright yellow, is also common in Kent (C. fragilis). There are many members of this family exceedingly minute in size, as the Pistil and Rush Clavarias (C. pistillata and juncea) which grow parasitic on dead straws and twigs.

The Calocera group resembles the last named; but the texture becomes horny when dry; hence the name Calocera. The orange species (C. cornea, *Plate XVIII., fig.* 15) is richly tinted. We have found it on stumps of oak and elm at Hawkhurst. The Typhula group also preserve the club shape; they are very minute plants, parasitic on dead leaves and stems.

The Pistillariæ have the same character, but are tougher in texture.

The last order of the hymenium family is that of the Tremellini, or jelly fungi. Here the fruit bearing part surrounds the whole of the plant.

Upon rails, fencing off water meadows in Wiltshire, we first saw the orange Tremella (T. mesenterica, *Plate XVIII., fig.* 16). It was later in the same season in which we had commenced our collection ; and when we saw puckered masses of jelly seeming to ooze from inequalities in the damp railing, we made sure we had got another kind of fungus. We watched the rails, and as long as the weather continued damp the jelly enlarged ; but it was not sticky to the touch as we expected it to be, a kind of skin covered it, keeping each lobe in shape. The substance was semi-transparent, orange at the base, and gradually growing lighter towards the summit. Shortly afterwards we found a large plant of the Foliate Tremella (T. folcasa) growing in the fork of an old tree at Monkton, Deveril. It was a rich claret colour, of an irregular shape ; and was very difficult to secure without destroying it. It seemed as if formed of port-wine jelly. The white Tremella (T. albida) is smaller than either of these, and less lobed. I have found it on branches and rails in damp woods in Kent.

The Exidia group vary in having the hymenium glandular.

The glandular Exidia (*Plate XVIII., fig.* 17.) somewhat resembles half a huge mulberry, or half a chequered plum. I found several specimens growing on a dead branch of oak, among dead leaves, in a swampy wood near Richmond, in Yorkshire. The Upper part of the fungus was black, chequered with dark grey ; the under part was olive, and covered with downy hairs. The plants trembled on the stem as I carried them; but I got them home in safety. The largest measured one and a half inch across. I have also found it in Herefordshire. No fungus answers so unsatisfactorily to its description, or looks so unlike itself in different stages, as the Jew's ear (Hirneola auricula judæ, *Plate XXIII.*, *fig.* 18). Berkeley describes it as concave, and others speak of it as cup-shaped. We first saw it in Wiltshire, adorning a leafless elder-bush in the early winter. The plants reminded us of large ears; as long, and thrice as broad, as those of a full sized lop-eared tame rabbit. The likeness to ears was perfect; there were the veins, the thin grizly texture and a soft velvety surface like mouse-skin. The colour varied from red to green, in beautifully blending shades. The following spring we found the same plant on budding elder; it was thick, smooth, semi-transparent, and only velvety underneath; the veins were there, but it seemed impossible that the substance should ever be rolled out to the extent and thinness of our friends of last year. It is one of these young specimens that are figured in the plate. Comparing my specimens, I concluded that I had thus procured typical representations of the two extremes of Jew's ear life; but last year I found a perfect forest of it upon an elder hedge near Ross, in Herefordshire; and, lo, specimens were there of every age, and in every state of preservation. Infant gelatinous plants were there, rising one above another; full-grown ones, still shelf-like, not cupped, not pendulous; old shrivelled ones were there, very little like ears; some blackened with age, some covered with moulds, some zoned with confervæ, but not one the least like my beautiful lop-eared specimens!

The Dacrymyces group are minute fungi-like pimples of jelly on dead wood, or the trunks of trees. The orange species (D. stillatus) is very common; a constant parasite on pine rails. It grows in great abundance near the railway station at Warwick. The pretty golden cup Dacrymyces (D. chrysocomus) is rare; we have no specimen of it.

Thus we reach the end of the first great family of fungi; those whose fruit-bearing part is exposed. We would fain clear away the prejudices against the whole race, but this is difficult; for in the present day prejudice can only be vanquished by utilitarianism, and we can only claim the protection of that for a very limited number of fungi. People have been too long accustomed to consider them as evil signs; and Shelley furthers the popular superstition.

> "Agarics, and fungi, with mildew and mould,
> Started like mist from the wet ground, cold,
> Pale, fleshy, as if the decaying dead
> With a spirit of growth had been animated."

Surely the high intelligence of the nineteenth century would do better to emulate Bishop Mant's style of dealing with the fungi! Let us prevail on our gentle readers to tread in his steps.

> "For mostly in the forest dank,
> Or mid the meadow's herbage rank,
> When flora's lovelier tribes give place,
> The mushrum's scorn'd but curious race.

Bestud the moist autumnal earth ;
A quick, but perishable birth.
Prompt to alter, fade, decay ;
Tho' much you fail not to admire,
Their parts, their structure, their attire,
The pillar stem, the table head,
As with a silken carpet spread,
Inlaid with many a brilliant dye
Of nature's high wrought tapestry ;
Of autumn's waning strength they speak,
And tell how nature, worn and weak,
Prepares her sceptre to resign,
And in inactive languor pine."

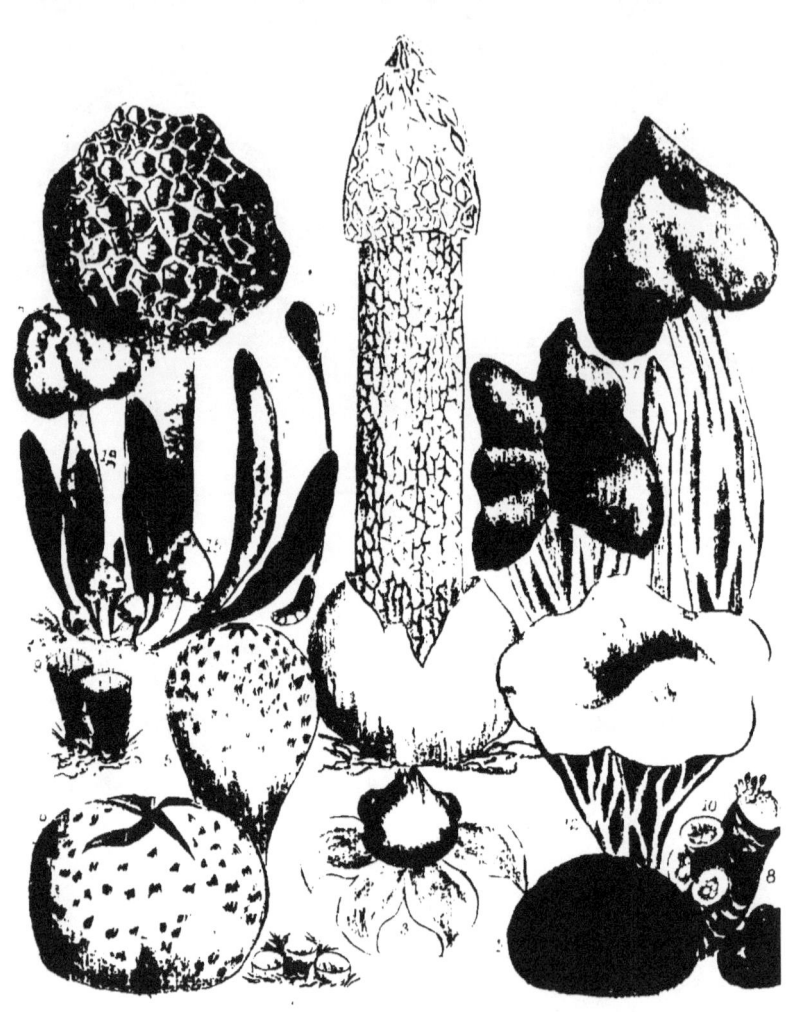

CHAPTER XXVII.

FUNGI.

"We'll make a feast in our mossy dell,
Of infant puff-ball and rare morel,
And many a favoured guest shall sup
On lily dew from a siller cup.
The aged puff-balls shall help us to cheat
The dainty bees of their luscious meat;
While others shall burn to give us light,
And scare from our dell the dreary night."

WICLIFFE LANE.

THE second class of fungi are characterized by having the hymenium, or fruit-bearing part, enclosed in a single or double envelope. It is therefore called the envelope class, or *Gasteromycetes*. In this class we look no more for the column and capital, though that form appears in exceptive cases. The general contour is spherical, or approaching to spherical, but there are many variations from this, and forms of great elegance and interest.

The first order of envelope fungi is subterranean in its habits; its characteristic distinction is that the hymenium does not turn to dust until the whole plant decays. This order is called Hypogei. When staying near Bath, my kind friend and instructor, Mr. Broome, one of the leading fungologists of the present day, took me out to search for plants of this order, one of which, the " Bath Truffle," used to be sold for the same purposes as the real Truffle in the market at Bath.

We went to the beech grove where he had often found specimens, and a dog trained to sent out the fungi accompanied us. But, alas, either the False Truffles were all gone, or the dog had lost his scent, for we moved the dead leaves, and scratched the surface of the earth over a large area, but no specimen rewarded our search. Mr. Broome gave me a specimen of the plant in question from his own herbarium (Melanogaster variegatus, Plate XIX., fig. 1). There are six groups of these subterranean fungi according to Berkeley, but we have been eminently unsuccessful in finding specimens.

The next order is that of Phallus. It is characterized by a perfect envelope, out of which the fruit-bearing part rises, and in a short time melts away. In the Common Phallus we have an envelope closely resembling that of our first friends, the Fly and Phallus-like Agarics. The young plant appears like a round puff-ball, but the envelope bursts, a tall honey-combed column shoots up, crowned by a cap in true Agaric style, which cap is at first covered with an olive coloured hymenium, gelatinous and moist, and which oozes away quickly, giving forth a most offensive odour, and attracting swarms of flies to the noxious feast. The plant has a dignified and imposing appearance, and might well be accounted a desirable ornament of woods and pleasure grounds, but for its abominable and all-pervading odour. This procures for it the name of Stinkhorn, but I greatly prefer the more euphonious one of Wood-witch. On one occasion, when, after a long absence from Yorkshire, I returned to visit the beautiful woods of Swaledale, my friend conducted me, as of yore, to my favourite spots in the wild and rocky thickets. Summer had just burst forth in its full loveliness, the foliage was thick, the grasses shook forth their flowering panicles, and I exclaimed that the lovely nooks had become more beautiful than ever. But my host assured me there was now a sad drawback to the charm of the near rocks, our favourite sylvan drawing-room of olden days. He said some rabbits or larger animals must have died in the underground caverns, for the rocks were so heaped together as to form subterranean grottos and passages, giving rise to a legend of a secret way under ground all the way to Easby Abbey. He said that these deceased animals gave forth a stench so disgusting, that they were chased from their favourite haunt, and the summer evenings could no more be spent among the near rocks. Still, I wished at once to revisit the familiar scene ; but as we approached the odour he complained of became painfully perceptible. Another turn in the path, and I beheld a group of Phallus, tall and stately, like a group of marble obelisks, fretted with elaborate carving and quaint devices. I laughed triumphantly, and hastened to assure my friend that the deceased animals were no other than living Stinkhorns ; and if he would make his gardener remove all such interlopers from his plantations, the accustomed nooks would be tempting as ever for evening lounges (Phallus impudicus, *Plate XIX., fig* 2).

The Clathrus is a handsome plant of the same order, and excells the Wood-witch in its abominable perfume. Our one English species (C. camellatus) is peculiar to the south of England and Ireland. It is red, and branched like coral or sealing wax.

The third order of the envelope class, that of *Trichogastres*, is characterized by a single or double envelope, enclosing the hymenium, which presently turns to dust and threads.

The group of Earth Stars (Geaster) is the prettiest of the Order. Here there are two strong envelopes, the first bursts early in the life of the plant, and immediately tears into equal segments, turning back, and lying upon the ground in a starry form. The second remains entire till the spores are ripe, then it opens in the centre, and allows them to escape. Mr. Broome gave me specimens of the Hairy Earth Star (G. fimbriatus, *Plate XIX., fig.* 3), which he found in great abundance about Bath.

The brown puff-balls, which we find in pastures so frequently, belong to the next group, Bovista (B. nigrescens, *Plate XIX., fig.* 4). We used to gather them as children, on Hungry Hill, near Ripon; and there is no neighbourhood, in which I have sojourned in autumn, where I have missed the familiar bag of dust. The Scotch call them Devil's snuff-boxes. There is a smaller species somewhat lead coloured, it is quite as common (B. plumbea).

The giant puff-ball is commonly found in the Swaledale pastures. It grows to a great size, and in its youth forms a wholesome and pleasant dish, compared by some to sweetbreads (Lycoperdon giganteum). This puff-ball is used to smother bees, or rather, by its fumes to induce intoxication; and while the bees continue in that state the honey can be taken, and the lives of the honest insects spared for future labour. On the subject of puff-balls, old Gerarde is kind enough to give us his opinion, at the same time introducing us to an ancient use of the plants. "Puff-balls are no way eaten, the powder of them doth bite. In divers parts of England, where people doth dwell farre from neighbours, they carry them kindled with fire, which lasteth long." Imagine an evening party assembling by the light of glowing puff-balls; it would rival a gypsy pic-nic in excitement! Our grandmothers, however, did use dry puff-balls as tinder, and very suitable they seem for natural tinder boxes.

The pear-shaped puff-ball (L. pyriforme, *Pate XIX.*, *fig.* 5) we found about tree stumps in the Sheerwater woods; it was furnished with a root.

The Scleroderma group has a double envelope, the outer beset with clothy scales. The common species frequents wood-borders, often growing in clusters; it is common enough; our specimens are from Hawkhurst (S. vulgare). The S. bovista I found in Cornwall; it was larger than the common species (*Plate XIX., fig.* 6) and the spores were yellow-olive. The smell was not pleasing, and I was sorely tempted to throw it away, lest it should damage my pretty basket of Cornish flowers. But fungi are scarce in July, and I was unwilling that Cornwall should be unrepresented in my fungus collection; so I placed my yellow puff-ball beside the red Broom-rape and Cornish heath, and pressed on towards the Lizard Point.

The fourth order of envelope fungi (Myxogastres) contains a comparatively uninteresting set of plants. In their youth they are soft, but presently they become a mass of threads and dusty spores. They are furnished with an envelope, which keeps the plant in shape till nearing maturity, when it tears and the dusty contents become diffused. Masses of yellow dust lying upon the sward, or umber dust, in similar situations, represent exploded individuals, or clusters of Reticularia, or Æthalium. A little Didymium, which I found in numbers on a dead leaf, belongs to this order. It looked like a miniature Agaric made of silver paper, and the envelope showed scales on its surface under the lens. The Arcyria group contains some pretty species. The Red Arcyria (A. punicea, *Plate XIX., fig.* 7) grows in abundance on dead wood. In Kent, Wilts, and Yorks I have found its clustering heads full of vermillion dust; the top of the envelope soon gives way and then the spores escape. In Trichia the envelope tears lengthways, and the species are without stems; they are common in hollow trees and under dead leaves (Trichia Turbinata, *Plate XIX., fig.* 8).

The fifth and last group of envelope fungi is called Nidulariaceæ, or bird's-nest fungi. It is possessed of marked characteristics, the spores being packed in various egg-like parcels, and the eggs enclosed in a tough or woody envelope, well compared to a bird's nest, and to which the eggs are attached by a spiral cord, which uncoils when the seeds are ripe, and jerks the little parcel out, as a bird shoves its timid or lazy brood into freedom and independence.

In the shady depths of Kingswood, near Congresbury in Somersetshire, while seeking mosses and ferns in the late autumn, we came upon a group of exquisite fungi—in form like the upper part of a wine glass, the rich brown of the envelope (peridium) was regularly striped, and covered with scattered hairs, which rose as a fringe round the edge of the cup. Many of the little parcels of seeds were already scattered among the beech mast and the gray lining was striped like the outside. The cup might have been large enough to fit on the end of a child's finger, but it was too small for ours. We considered our striped Bird's-nest a great treasure (Cyathus Striatus, *Plate XIX., fig.* 9).

The neighbourhood of Edinburgh furnished us with the Bell-Bird's nest fungus (C. vernicosus, *Plate XIX., fig.* 10). Dr. Greville describes it as frequenting "neglected gardens;" but the garden where our specimens were found, did not by any means deserve this appellation. The beds were gay with successive flowers from the "Fair maids of February" till the Michaelmas daisies. Not a dead leaf was allowed to rest there, not a straw suffered to remain out of its place. Yet upon the box edges appeared the horny cups of the Bell-bird's-nest, pallid externally, white within; its seed parcels carefully packed, and waiting for delivery; its every manner and habit harmonising with the neatness and regularity of its home. In Scotland it is called "Sillercup." In these two species of Nidularia the elastic cords attached to the *sides* of the envelope.

In the Crucibulum group the attachment is to the bottom of the cup; and herein consists the one distinction between these bird-nest groups.

The Crucible Bird's nest (C. vulgare, *Plate XIX., fig.* 11) partakes of the form of the vessel from which it is named; not tapering to the base as its two cousins do.

There is a long narrow field near Kemberton, in Shropshire, with lines of trees on either hand. It looks like a grand approach to some noble mansion; but it leads nowhere, and has no attraction beyond its own beauty; pasture fields, or arable fields are around it; but it is the most attractive nook amongst them all. Here the wood anemones quiver their pale heads in hundreds; primroses peep from under the hedges; and violets nestle about the tree roots. Here mosses flourish in endless variety, sheltered from the midsummer glare by tall grasses, and spreading ferns. And here, when the grasses are withered and the ferns are dead, clusters of Crucible bird's nests spring from the decaying stems, closely guarding their fruitful eggs until the moment when the cord is bid to jerk, and the fungus seed mingles with the fresh leaf mould, and prepares to bide its time till the spring flowers and summer grasses have had their turn, and the little crucibles will be due again.

A very scarce fungus belonging to this order rewarded my search in Yorkshire, one winter's day. Passing the Richmond race-course, the main-road led through a dark wood; here I left the public path, and betook myself to the wood, searching for mosses and fungi. While examining the ground under some fir trees for varieties of the fork-moss family, I saw some orange bodies, like beads, about the size of rape-seed. These I took, concluding them to be fungi. The next post carried them to Bath; and in a day or two I received a request from Mr. Broome for more of the orange beads ; they being plants of the ground Thelobus, a fungus not recently found in Britain. Of course I was proud in the extreme ; but for many days illness and stress of weather prevented my repairing to the fir wood. Snow came, and botanical interests were necessarily held in abeyance ; the snow lay long, but when softer airs had melted it even from the black fir wood, I searched again, but not one trace of the precious orange beads could I find!

CHAPTER XXVII.

FUNGI.

"Love Nature, and her smallest atoms
Shall whisper to thy mind."
CHARLES MACKAY.

HAVING examined the principal members of the Hymenium and Envelope classes, we now come to the third or dust class, Coniomycetes. Here the spores are the leading feature; sometimes they are single, sometimes connected in chains, sometimes covered, sometimes lying in naked clusters, and supported on more or less rudimentary threads; but whatever the variety of the case the minute plant seems chiefly composed of dust-like spores.

The large majority in this class are epiphytes, or parasites, according as they infest leaves or wood, and other substances. All are minute, and are only recognisable by means of a lens.

Wherever we find a bundle of dead nettles we may hope to discover black specks on the stems; which, when examined by means of the microscope, are seen to be bottle-shaped. This nettle dust is one of the Aposphæria group. Sedge leaves furnish an allied epiphyte, but the substance of the fungus is mainly hidden, and only the mouth appears on the surface of the leaf; this is the Sedge Neottisporia. Decaying holly leaves are found marked with a circular spot, very pretty and attractive; this is another allied species, Ceuthosporia Phacidioides. In the woods about Richmond, infected leaves are found in great abundance.

The simple Torula appears like a mere stain of soot on tree stumps; it is of very frequent occurrence in woods where trees have been recently felled. Bramble leaves afford pretty specimens of Aregma looking like minute rusty spots to the naked eye, but showing pretty clusters of stems, each surmounted by an oval head under the microscope. This Aregma belongs to the Puccinia order, which contains a large number of pretty and of dangerous species.

When gathering spikes of betony for a noosegay, I was attracted by the phenomenon of apparent fern-seed upon the leaves. The under surface of almost every leaf upon the plant was dotted with brown clusters of spores, closely resembling the seed masses on the back of the shield ferns. This proved to be extensive plantations of Puccinia Betonica; I preserved some specimens then, and having revisited that Kentish lane two successive seasons, I still find the betony leaves laden with Puccinia. In Swaledale we have found dandelion leaves similarly ornamented, but chiefly on the upper surface; here the clusters are smaller, and less swollen; and never become confluent (P. variabilis). In the same neighbourhood the Devil's-bit Scabions appears attended by its especial epiphyte, another of this Puccinia family; the Sanicle, likewise, has its humble dependant of the same order; the Adoxa its peculiar fungus; while here and elsewhere the bean, the box-tree, the figwort, the periwinkle, the golden rod, the willow-herb, and the enchanter's nightshade, are similarly adorned.

The Uredo group are orange-brown, or black. A few members of brighter tint occasionally appear. I found a yellow Uredo on the dog's mercury, at Hawkhurst; and a handsome orange one on the strawberry-like Cinquefoil, in the same and other districts; the willow herb, St. John's wort, oak and bilberry, each nourish one of these gayer Uredos.

At Clevedon in Somerset, we found Sea-lavender whose leaves adorned with brown specks, furnished specimens of one of the Brown Uredos; and in the Yorks woods we have gathered leaves of the Herb Robert, studded with brown dots as fine as a needle point, a close ally of the Sea-lavender Uredo.

Some Maiden-blush roses, growing in a cottage garden in Swaledale, had leaves half-covered with yellow dots; those under the microscope appeared clusters of spores, bursting from the mother cell. The elegant little Cathartic Flax had similar specks on its leaves, and we soon ascertained them to be an allied species of Epiphyte (Lecythea lini). When traversing lanes in the neighbourhood of Edinburgh, early in one spring, we noticed plants on the Wayside Shepherd's Purse all blotched, as it were with whitewash. We concluded that some had been used to the adjacent wall, or at any rate carried along the street. But as we left the neighbourhood of houses, and the lane became a real country lane, we still beheld the marks of the whitewash! On close examination we found that this was a white epiphyte to which the Shepherd's Purse was subject.

Leaves of the Water Sweet grass growing beside certain ponds at Hawkhurst, are marked with long sooty lines accompanying the veins of the leaf, this is the Elongated Ustilago, and is a fungus of decidedly bad character. It propagates so quickly that one plant of it, so small as to be invisible to the naked eye, will bring forth forty seeds. These seeds float in the air, and soon find homes on the stems and leaves of corn and grasses. When first this blight appears, it is orange in colour, but presently turns a rich brown, and then becomes black. This introduces us into that evil group, the St. Gileses of the fungus kingdom, where the dangerous members of the dust class are to be found. Here is Smut (U. sagetum), which takes its rise within the glume of living plants, and grows with such rapidity as speedily to fill the interior, and burst through the skin, showing itself as dirty black dust oozing forth. Withering describes this as consisting "of very minute, egg-shaped stemless capsules, at first white, but the thin white soon bursting, it pours out a quantity of brown-black powder, mixed with wool-like fibres."

Bunt (U. caries), is the plague of wheat, as Smut is of other cereals. It not only destroys the ear on which it grows, but every grain that comes in contact afterwards

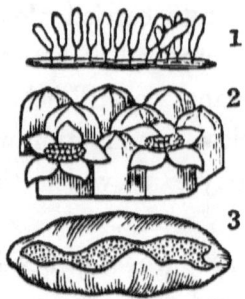

1. WHEAT BLIGHT.
2. GOOSEBERRY BLIGHT.
3. BUNT.
All magnified.

with the infected individuals. The spores when crushed give forth a disagreeable smell, and bunted wheat used to be applied almost entirely to the making of gingerbread, where the other condiments employed disguised the unpleasant odour.

The Buck-wheat has its smut (U. utriculosa), the Sage its smut (U. urceolarum), and other plants are similarly endowed.

Lest the reader should turn with disgust from the whole class of Dust fungi because of these harmful Uredos, we will introduce him to another member of the class which we should have spoken of before had we not wished to reserve it as a bonne bouche after the Smut and Bunt. The Yeast plant (Torula cervisea, *cut*), consists of round or oval cells; these cells at first are solitary, but within an hour of being placed in a good habitat, other buds and cells have appeared; in three hours these are doubled; in eight hours branching cells appear, then the mature cells explode, giving birth to numerous young cells, and in three days threads and branches are produced. Hogg, from whose work I have taken this description, also mentions another stage of the plant prevailing in porter vats, but as in this stage it is not beneficial, but rather the contrary, we will not enter into the elaborate discussion of it. But surely the widespread utility of the Yeast plant, from which even the temperance society cannot withhold a fair meed of praise on account of its effect upon our bread, will go far to redeem the Dust class from disgrace. German yeast is the cells of the Yeast plant in a dried state.

If it lies with the Yeast plant to substantiate a claim for utility in its class, the Æcidiums may well establish one for beauty. Mr. Perry of Warwick first introduced me to the beautiful circles of Æcidium plants adorning the leaves of the primrose (Æ. primula). The fungus abounds in the neighbourhood of Warwick, but is not frequently found in other districts. Each little plant in this group is enveloped in a thin covering, which opens in regular segments like the Earth Star. Thus, while the naked eye only discovers a dot on the surface of the leaf, the lens reveals a bright cluster, each pleiad of perfect form, shading from an orange centre to the transparent white which tips the rays. In Yorkshire we found the Wood anemone leaves spangled with Æcidia, some curled back as if the weight behind contracted them. Mr. Perry gave me leaves of the Garden anemone similarly peopled. Woods in Yorkshire afforded abundant specimens of the very pretty Barberry Æcidium infecting the leaves of the shrub of that name with its bright orange blotches. In former days the presence of this epiphyte brought the Barberry into evil repute, for the farmers associated the yellow blotches with the orange lines on the leaves of cereals which ended in the Blight. They used therefore to exterminate the Barberry from the hedgerows, lest its parasites should infect the corn. Gooseberry bushes often bear a large crop of Æcidium speckled all over their leaves. Here the individual plants are crowded together in thick patches, and when each has opened its starry lid the blotch looks like honeycomb under a lens. The Mountain Willow-herb, so abundant in the York woods, is often dotted with especial Æcidium, and in this species the plants grow singly. The Wood Spurge covering the Herefordshire hills with its sunny foliage, exhibits another species of Æcidium, the especial retainer of the Spurge family. Surely when we examine these plants, so marvellous in the beauty of their structure, the extreme minuteness of their size, and the inconceivable number of their individuals, we shall no longer despise Dust fungi, but rather say with the poet,

> " God made us, as he made things all,
> In perfect beauty, tho' so small.
> Take a lens, you'll see we're not
> Despised as blight or Mildew spot,
> But Smut and Canker find a place
> Amid the varied fungus race."

The class succeeding that of the Dust fungi, is characterized by the predominance of *threads*, and is therefore called *Hyphomycetes* or *Thread fungi*.

The plants of the first order, (Isariacei), are parasitic on twigs, dead flowers, fungi, and insects. Other groups follow their species, forming dark coloured specks on decaying stems, bark, and leaves, presenting little variety of interest except when examined with the microscope.

The order of Moulds (Mucedines) have interest enough, both from the extreme beauty of their structure, and the important part they play in the life of man.

And first we will consider that fungus of gastronomical celebrity, known as Blue-mould in cheese (Aspergillus glaucus). Rising on a crystal stem formed of a single cell, chains of cells radiate from the summit in great numbers making the plant into a miniature standard bush. Yet to the naked eye a forest of these bushes appear but as a blue stain!

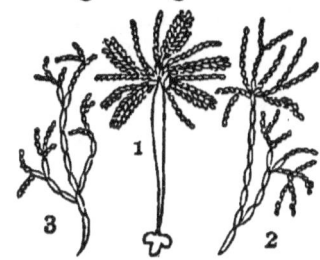

1. BLUE CHEESE-MOULD.
2. HERB-MOULD.
3. DUNG-MOULD.
All magnified.

Another Aspergillus grows upon the plants in herbariums, but its stem is formed of many cells, and the chains of the shrub are repeatedly branched. Other species appear on dog and rabbit-droppings, and present the same shrub-like structure. Aspergillus is the name of the brush with which the holy water is sprinkled in Roman Catholic churches, and is hence adopted for these dainty plants.

The Botrytis group has its infamous member, the Potatoe Blight (B. infestans). The threads of the mycelium enter the stems and leaves, and prevent the circulation of the juices, while a villanous ally (Fusisporium solani) attacks the tuber, and spreads in every direction, consuming the grains of starch with which the cells are filled, and supplying their place with their own noxious branches. The Tomato is infested with the same unwelcome parasite, growing from the root or spawn, throwing out branches, and nourishing itself at the expense of the infected plant. Another Botrytis preys on fruit, and all decaying vegetable structures. One species attacks the living silkworm, while the house-fly and other insects are preyed upon by other members of the order.

The Oideum group contains some important members. The Mould of the pear and of the orange (O. fructigenum and fastigiatum) and that of the grape are well known. The last named is thus described by Mr. Harris:—
"Grapes, when blighted, are covered with what appears to be a white powder, like lime, a little darkened with brown or yellow. This powder, being a collection of fungi, sends forth laterally, in all directions, thread-like filaments, which become so completely interwoven with one another as entirely to cover and enclose the skin of the grape in a compact and firm network, and on each is seen the egg-shaped capsule or seed-pod."

Recent discoveries in medicine and anatomy have convinced physicians that fungi bear a considerable share in the diseases which attack the human frame. Ringworm is caused by the germination of a fungus upon the skin (Oideum porriginis); and during the time of the cholera visitation in 1854, the Rev. G. Osborne collected many microscopic fungi from the air, and smaller ones have been detected arising from foul drains.

The vinegar plant is also a thread fungus (Mycoderma aceti). Mr. Slack describes it as "a tough leathery moss, often used by private families to make vinegar out of solutions of sugar and treacle. If a thin piece of the large, tough vinegar plant is examined microscopically, a moderate power suffices to show an unorganized jelly and cellular structure of many shapes, often resembling coherent cells of yeast, others being like Oideum. It is also, in those I have examined, easy to see something like an entangled mass of minute threads.

This plant is formed in vinegar obtained from wine or beer, but not in that obtained from wood.

Such are the fungi composing the two classes of Dust and Thread parasites (Coniomycetes and Hyphomycetes). They are all-pervading, existing in chemical solutions, in tropical and polar climates, vegetating on glass, iron and horn, as well as wood and foliage, and studding the silks and gloves in a lady's wardrobe as freely as the dead straws by the pool side. Moulds are even traced in fossil wood.

CHAPTER XXIX.

FUNGI.

> " Various, as beauteous, nature, is thy face ;
> . . . . . All that grows, has grace.
> All are appropriate. Bog, and marsh, and fen
> Are only poor to undiscerning men.
> Here may the nice and curious eye explore,
> How nature's hand adorns the rushy moor ;
> Beauties are these that from the view retire,
> But will repay th' attention they require."
>
> CRABBE.

THE first large division of the fungus tribe was characterised by naked seeds or spores. In the Hymenium class these were contained on an exposed surface, in the Envelope class they were enclosed in an envelope or case, in the Dust class they formed the main part of the plant, and in the Thread class they were merely attached to the thread-like stem or branches; but greatly as the different classes varied, they were unanimous in the simplicity of their spores. We now come to the second great division of fungi, where the spores are contained in bags or *asci*, hence the first class in the division is called Ascomycetes or Bag fungi.

Plate 20

1 Scarlet Peziza 2 Orange P 3 H...bed P. 4 Green P Veined P 6 Claret
5 Black & rusty P. 9 Verdigris P 10 Hoary P 11 Ground P 12 Bulgaria. 13 Sh..p..ess
Hypoxilon. 14 Candle snuff Xylaria. 15 Clumsy Xylaria

In the Elvellacei group the Bags are disposed on a Hymenium nearly as much exposed as in the Agarics and Boleti.

The Edible Morel comes first in the order of arrangement (Morchella esculenta, *Plate XIX., fig.* 13). This fungus rises upon a columnar stem, and its head is round or oval. The Hymenium is spread over the surface of the cap, and is honey-combed into cells. We have gathered it frequently in groves and wood borders, and the sight of it brings to our memory the fresh air from the hills, and the ripple of the peat-stained stream over the fine limestone rocks, and the cry of corn-crake, and the rustling of young foliage, and all the charms which make the Yorkshire dales such a land of delight. We never passed the Morels, for we loved to propitiate the good graces of the cook by a timely gift of them, to flavour her gravies with ; she had always a string of them hung in her store room, and she was ever in fear lest her stock should be exhausted. No fear of incurring her displeasure, even for keeping dinner waiting, if we brought a handful of Morels, " Jews' cars again !" she would say, handling the plants affectionately, " well, it is a good thing when folks keeps their eyes open." But the Germans in olden days were even more alive to the gastronomical excellence of the Morel than our Yorkshire cook. It was found that the fungus flourished in the greatest luxuriance on wood ashes, and such was the rage for the delicacy that large portions of the forest used to be burned down annually, on purpose to secure a large crop of Morels. The practice was stopped at last by an express order from government.

Once only we found the Loose Morel (M. semilibera) in a thicket on the banks of the Swale. Here the stem is free from juncture with the substance of the head for some distance, though overshadowed by it. The honeycomb is larger, and the head much smaller in proportion to the height of the stem, than in the edible species. The Spread Morel we have none of us found.

The Mitre Helvella (H. Lacunosa, *Plate XIX., fig.* 14) was one of our highland treasures. We had spent the night at Ballahulish, and as the steam boat would not call on its way to Fort William until noon, we set forth for an early ramble, hoping to gain treasures for our botanical collection. We rambled up the hills, crossing runlets, and stony places, where waters raged in the winter, and finding mosses in profusion, and beautiful lichens, and delicate branches of the ramping fumitory, and heath in abundance. But no fungus greeted our eyes till we reached the low ground again, and the distant steam of the boat appeared cloud-like over the waters. Then, with an exclamation of delight, we sprang forward to gather the most curious and uncanny group of fungi that we ever beheld. The stems were all wrinkled and grooved, the heads bloated and bulged into every possible form. One might have been an old mitre made of felt, and knocked about and weather-beaten, till it was all bulges and hollows, another was more like a cocked hat, while others again were club-shaped, only verging slightly towards the form of the mitre. The black heads contrasted strongly with the white stems, and the effect of both was heightened by the lovely moss carpeting the ground beneath them, and by some plants of sheep's sorrel which leaned their crimson leaves against the grooved columns. In this group the head or *receptacle* hangs down over the stem, and the fruit-bearing surface is on the upper part.

In the succeeding group of Verpa, the receptacle is bell-shaped; and rutted on the outer surface. We have not found either of the species.

The Mitrula group have oval receptacles. These are scarce fungi, the only time we found either of them, was when accompanying a shooting party on the Yorkshire moors. There, where the oozy ground warned us to beware of peat bogs, we espied among the spagnum and marsh plants, the orange heads and white stems of the Marsh Mitrula (*Plate XIX., fig.* 15).

The Spathularia is distinguished from the Mitrula by its long narrow head, flattened, and bright yellow in colour. Once only I have found it, growing under fir trees in the Fright wood near Hawkhurst (S. flavida, *Plate XIX., fig.* 17). The same neighbourhood furnished our specimens of the shining Leotia (L. lubrica, *Plate XIX., fig.* 16). There is a lane which has yielded me many a floral treasure, where the primroses grow in galaxies, and the Lent lilies nod by hundreds in the breeze, where the spotted orchis lifts its noblest spikes, and the Lady fern grows in greatest variety of form; there, on the high banks of that sheltered lane, hard by the roots of the now sleeping flowers, rose the pale stems of the Shining Leotia, upon which the glossy heads of bright olive, puffed and swelled to their utmost.

The next group, that of Geoglossum, takes me back to

Scotland again, across the wild road leading from Loch Lomond to Oban. The turf is short, and few flowers vary its hue in September; but amid the verdant pennons of the grass rise thick black tongues, shaped like simple clubs, and looking like imps turned to wood by the light of day. This proved to be the Hairy Geoglossum (G. hirsutum). At first sight I believed it to be a clavaria, but the spores being contained in bags at once distinguished it from that family.

The next Geoglossum added to our collection was the olive species (G. olivaceum, *Plate XIX., fig.* 18). It was growing on the moors above Oban, hard by the Amethyst Clavaria. Then came a box of Geoglossum, black as the impish hairy species, but as large as the olive one, and all glossy with moisture. This was the Shining Geoglossum (G. glabra, *Plate XIX., fig.* 19). It was sent to me by a sister botanist in Wiltshire, who had found its black tongues looking strangely weird beneath stately trees in a gentleman's grounds. I afterwards found the species in woods in Herefordshire.

We now come to a family unsurpassed for beauty and charm among the fungus peoples, though having no claims to utility. The Pezizæ are cup-, or, in some instances, saucer-shaped; the hymenium spreads over the inside of the cup. The spores are amazingly small and light; and if, when they are ripe, you irritate the surface of the cup with a feather, the spore-bags burst, and the spores rise like a puff of smoke.

We were rambling in my favourite Yorkshire woods one fine spring day, and recent rains had made the steep path very slippery—nay, worse than slippery: the mud lay so thick that, while it refused to give us foot-hold, it nefariously robbed us of our goloshes. We were thus stranded on a grassy bank, and angling with our umbrellas for the mud-locked goloshes, when my eye fell on a large mis-shapen brown cup, pale on the outside, and wrinkled within. It was as large as a tea-cup. I was at once reconciled to my difficult position, and felt quite obliged to the mud for staying my footsteps, and giving me time to distinguish the clay-coloured cup from the clay out of which it sprung. This was the Veined Peziza (P. venosa, *Plate XX.*, *fig.* 5).

Our first specimens of the Orange Peziza were found in Kent (P. aurantia, *Plate XX.*, *fig.* 2). Here the cups are nearly as large as in the last species. The outside is pale orange; the inside most intense and brilliant in colour. The plants grow in clusters, and nearly always shoulder each other out of shape. It was growing on walks in the pleasure grounds both at Risden and Elfords one autumn; and the next season I espied its orange cups in a far distant and far different place, the cemetery at Glasgow.

The Hot-bed Peziza is an attractive species from the extreme neatness of its globular form, and the umber scales which often stud its surface. Our specimens are some from Kent, and some from Shropshire (P. vesiculosa, *Plate XX.*, *fig.* 3).

The Pale Peziza is smaller than any of these, and somewhat boat shaped. It is whitish and scaly on the outside, and buff within. I gathered it off a stump on the banks of Loch Lomond.

The Scarlet Peziza is deservedly a universal favourite.

One graceful writer calls it Fairy Bath; and I ever think Bryant's lines most suitably apply to it:—

> "Scarlet tufts
> Are glowing in the green, like flakes of fire,
> And wanderers in the prairie know them well,
> And call that brilliant plant the Painted Cup.
> * * * * * *
> * * * These bright chalices were tinted thus
> To hold the dew for fairies, when they met
> On moonlight evenings in the hazel bowers,
> And danced till they were thirsty.
> * * * * * *
> Well,
> Let then the gentle Manitou of flowers,
> Lingering amid the blooming waste he loves,
> Tho' all his swarthy worshippers are gone—
> Slender and small, his rounded cheek all brown
> And ruddy with the sunshine; let him come
> On summer mornings, when the blossoms wake,
> And part with little hands the spiky grass;
> And touching, with his cherry lips, the edge
> Of these bright beakers, drain the gathered dew."

The Scarlet Peziza (P. coccinea, *Plate XX., fig.* 1) grows on hazel twigs. Insects and slugs are very fond of it, and it is rare to find a full-grown specimen free from their dilapidations. It appears generally about January, and, with moss, makes a charming winter bouquet, especially if a few snowdrops or a Christmas rose can be had to heighten the force of contrast. We have gathered these beautiful fungi in Shropshire, Herefordshire, Wiltshire, Kent, and Yorkshire.

A very small Grey Peziza, frequently found on the ground

or on twigs lying on the ground, rewarded me for weeding a friend's fernery in Kent. The pretty fungus was nestling beneath the ferns in great numbers, its cups wide spread, about half an inch in circumference, in texture waxy, and fringed round the edge and underneath with brown hairs. The tiny white Peziza (P. virginia) is found in quantities parasitic on dead twigs and moss; its cups are only the size of a pin's head, and its stem is short and thread like. The banks of Loch Lomond furnished me with a beautiful claret Peziza; no description can depict the extreme depth and richness of its colouring. It grew on the bare ground (*Plate XX. fig.* 6).

The green Peziza (P. vermiformis, *Plate XX., fig.* 4) grew on old willow stumps, in a pretty field hard by a stream near Bath. It is a rare species. The bright yellow cups of the tiny P. Claro-flavum were found on a decaying log in Wiltshire (*Plate XX., fig.* 7). The same neighbourhood furnished the rusty black Peziza (P. atro-fusca, *fig.* 8) and the red-brown shields of the hairy Peziza (P. hirta, *fig.* 10). Last autumn, rain fell day by day, and as we looked forth on the dripping world, we longed that the rain would cease, if only to let us look upon the rich Kentish landscape; but the rain would not cease. Then, as if to afford us entertainment, brown specks appeared on the walk immediately underneath the windows, they enlarged into head-like bodies, these opened and showed dingy cups, which spread and spread till they might have furnished plates for a doll's dinner party. When we came out again the walks were peopled with hundreds of the ground Peziza (*Plate XX., fig.* 11).

In the woods of Kent and Sussex we sought in vain for the Verdigris Peziza. It is used by the workers in Tunbridge ware, to furnish the green tint used in their pretty patchwork articles; and it seemed natural that it should grow in the neighbourhood where it is in so much request. But the pleasure of finding the little Peziza and the spawn stained wood, was reserved for a Scotch ramble; when, after wandering in the Braid valley, and filling our basket with mosses, and agarics, and hoary lichens, an old rail met our view, all coloured with the verdigris tint we had so long sought in vain (P. æruginosa, *Plate XX., fig.* 12).

In that same neighbourhood, within sight of Braid Hill, is an old family mansion, called Craig House. The plantations surrounding this ancient dwelling give shelter to numerous fungi; and there I had the good fortune to find the lilac Bulgaria (P. sarcoides, *Plate XX., fig.* 12), belonging to a group closely allied to that of Peziza.

The next order to the Elvellacei is the Tuberacei, and here we come to subterranean fungi again. The external form of the true Truffles resembles that of the false Truffles as closely as the Geoglossum resembles a simple Clovaria; in both instances the seeds being simple or enclosed in asci is the only sure distinction. The common Truffle of our markets (Tuber æstivum) has a rough surface, it is found in the south of England, especially in Hants, under beech trees. On one or two occasions we have searched for these rare fungi, raking the ground in spots reputed to be affected by them, but without success. One friend of mine, an eminent Fungologist, has found them repeatedly, but he goes on excursions on purpose to search for them. On such occasions he carries a rake, and this attracts the notice of the idle boys who come in his way; and the result is often that he proceeds with an extensive guard of honour. Such a guard is trying in the extreme to a man who wishes to concentrate his attention on one particular search, and it tried my distinguished friend accordingly. The next day he prepared his defence, filling his pockets with apples; then, when his tormentors had gathered in good numbers, he flung the apples far and wide, and while they sought and scuffled for the tempting fruit, he fled and eluded all pursuit.

Many attempts have been made to cultivate truffles, but no great success has attended them. A number were planted in a beautiful deep dell in Somersetshire, called Goblin Combe; but we sought in vain afterwards, no harvest seemed to result from the seed-time. The best plan of sowing it is to steep ripe truffles in water, and when they have rotted there, pour the water over the ground. These fungi are in great request in the Italian markets; one kind commanding the price of nine shillings per pound. Another kind which is esteemed there has a strong flavour of garlic. Strasburg pie is greatly indebted to truffles for its excellence, and they enter largely into the composition of many famous dishes on the Continent. The Red Truffle (T. rufum) is found near Bath, but it is not an edible species.

The Phacidiacei order is the third in the Ascus class.

Most persons have remarked the black blisters which constantly appear on sycamore trees in the autumn, this is a mould belonging to the group just mentioned; it only attains maturity when the leaves have lain long on the wet ground. A similar mould infests the leaves of willow, and another the stems of nettle. A group of line shaped or oval plants, like mere discoloured seams on dead branches, is called Hysterium: many of the species are very common, none interesting except to the eager fungologist.

The third order of the Ascus fungi is one which contains a very great number of species, *Sphæriacei*; here the asci are enclosed beneath a woody or membranous covering: a few are of considerable size, but a vast number are microscopic.

The first group, Cordiceps, has fleshy heads and elongated stems.

The Soldier Cordiceps (C. militaris, *Plate XIX.*, *fig.* 20) is the prettiest of the family; it grows parasitic upon dead pupæ. I found it in woods in Herefordshire, and Kent, and it has been sent to us from the Yorkshire moors. Several other species are parasitic on sedge, furze, reeds, and larvæ; and one, the terrible parasite of cereals, may claim to be the most dreadful of the fungus poisons.

All cereals are subject to this Ergot (C. purpurea); but it most frequently attacks the rye. Flour made of the ergotted grain is unwholesome; and if taken in quantity, causes gangrene. One season a large proportion of the rye grown along the coast of Normandy was ergotted, and the poor being in a very great measure dependent upon this grain for subsistence, were obliged to eat of it. Frightful diseases ensued, the sufferers died in agonies, the limbs of many dropping off from sheer decay before death put an end to their misery. For some time the scourge was referred to supernatural causes; but at last suspicion fell upon the flour. Its effect was tried upon a dog, and the result proved the direction of the evil. But in the hands of science the fearful power of this plant is turned to good, and a very valuable medicine produced from it.

There is a mysterious poison used by the gypsies, and avowedly extracted from fungi. It is more than probable that it is an ally of the Ergot. A case of poisoning by its means occurred some three years ago; and a strict examination into the poison began, but stopped suddenly; at any rate the public were favoured with no more particulars. I subjoin an extract from one of the daily papers at the time; it conveys all the information I have been able to gather. George Barrow's works confirm the fact that such a poison is in use among the gypsies, and that it is of fungus origin; but they throw no further light upon it.

"Among other jealously guarded secrets of the gypsy race is the art of preparing what they term the 'drei,' or 'dri,' a most deadly and insidious destructive agent, and for which medical science knows no antidote. Analysis detects no noxious properties whatever; and the most careful examination, microscopical or otherwise, shows it simply to consist of apparently harmless vegetable matter. The 'drei,' then, is merely a brown powder, obtained from a certain species of fungus forming the nearest connecting link between the animal and vegetable kingdoms, the powder consisting of an infinity of sporules. These fungoid sporules possess the peculiar property of being farther developed only by intimate contact with living animal matter (as when swallowed, &c.); they then throw out innumerable greenish yellow fibres about twelve or eighteen inches in length. When the 'drei' is administered, usually in some warm drink, these sporules are swallowed, attach themselves to the mucous membrane, germinate, throw out millions of these silky fibres, which grow with awful rapidity, first producing symptoms of hectic fever, then cough, eventually accompanied by incessant spitting of blood, till death finally inevitably supervenes, usually in about a fortnight or three weeks' time. A case of this description came under my notice in Italy, in 1860. Although the patient was attended by eminent physicians, accustomed to deal with cases of slow poisoning, no suspicions of foul play were entertained till the day after the decease, when, an autopsy being held, revealed the cause of death. The fibres, the growth of which had ceased with the cessation of the animal life and heat that had supported them, were already partially decomposed; had another day or two elapsed no trace would have been left of the foul deed. If the analysis of the mixture in question reveal no deleterious drug, let a dog or other animal be daily dozed, as the gypsy recommended, with 'three drops' in some warm vehicle. The result would show whether the brown powder is, or is not, the world-famous and destructive 'drei.'"

A group of minute epiphytes succeeds that of Cordiceps. The Clematis draping the Herefordshire lanes is speckled with one Hypocrea, and the culm of the Fescue grass wears a broad belt of another, this latter being of a bright orange, turns the stem round which it grows into a perfect club, so as to earn for itself and its support a title of relationship with the reed mace (H. Typhina). Our specimens grew in Kentish pasture.

The Xylaria group boast some sizeable members; they are endowed with stems, and are covered with a dark woody coat. The clumsy Xylaria (X. polymorpha, *Plate XX., fig.* 15) is like a massive club. It is black and heavy, and looks like charred fragments of wood as it clusters on the decaying stump. We found it in great quantity near Maiden Bradley, in Wiltshire; and as we examined it more closely, we soon began to admire it. Upon breaking away the external crust the hollow interior of the plant became exposed to view; and not the dome of St. Paul's, when filled with the well assorted bands of charity children, could present a more perfect picture of order and regularity than did the graduating tiers of asci, so symmetrically arranged round the cupola of the Xylaria.

The Candle-snuff Xylaria (X. Hypoxilon, *Plate XX., fig.* 14) is common everywhere, and would be generally familiar but that a great proportion of the eyes that rest upon it mistake it for charred stems of heath or bilberry, and give no thought to the fact that as no other mark of fire is visible, it may admit a doubt whether heath can have been burnt there at all. The stem of this fungus is sometimes simple, and sometimes branched, always furry to the base, and generally wearing white powder on its tips. The seed is borne packed in tiers within the branches.

The Hypoxilon group has a few members well worthy of a passing notice. Large swollen blisters, hard, and brick red, growing on bark, presents the shapeless Hypoxilon (H. multiforme, *fig.* 13). Raised knobs, like the heads of rusty nails, studding hazel branches, are plants of the brown Hypoxilon (H. fusca) ; and carmine spots on the bark of beech shew clusters of red Hypoxilon (H. coccineum). Numberless stains and dots of every shade of brown, some thick, some hardly raised, some only discernible at all by means of a lens, mark a portion of the members of the Sphæria group.

Next to the Sphæria group comes that of Perisporiacei, interesting as containing the race of Mildews, so tormenting to the farmer and nursery gardener. The fruit-tree mildew is trying enough, the rose mildew is an unwholesome guest, but all vexatious parasites melt into nothingness before the bête noir of Hop-growers, all whose summer days pass in terror, lest the caterpillar eat, or the mildew choke their golden hopes centered in their beautiful hop vines.

The last group in the Ascus class is that of Onygei, and one of its curious members graces our collection.

Once more memory returns to a Yorkshire wood. It is not the season when poets rave about nature's loveliness. The spurge laurel on the clink bank is green, but its scented flowers are gone, the autumn flowers lie crushed and saturated with winter rain, and the river roars turbulently below, missing the sunshine which should gild its waves. Here is a steep footpath, but none care to traverse it now that nuts are out of the range of possibility. But when seeking fungi one ever chooses the most out of the way and unfrequented path ; and so I climbed in the footsteps of last autumn's nutters. Presently I espied an old torn felt hat, and upon it a fine crop of Onygena. This fungus resembles a Leotia in form, but it is very minute, not exceeding a quarter of an inch in height. The seed bags are contained within the head. All the members of this group flourish on decaying animal mater—one grows upon hoofs, and another on old bones. They are scarce, and my specimens were accounted great treasures. (Onygena piligina).

The second and last class belonging to the ascus division is that of Physomycetes, a family nearly allied to the Hyphocmycestes, except for these having simple spores, and those spores in bags or asci.

The Ascophora Mucedo is the fungus called Bread mould. If you subject a little of this to the microscope, you see a grove of tall-stalks, each with a bead-like head. While young these are of a milk-white colour, but presently they turn yellow; later you may see the spore bundles under the skin of the head. In another day or two the fungi begin to get darker, and presently the skin bursts, and the spores are scattered in the air. This mould is very unwholesome, and has been known to produce serious illness.

A red mould, appearing sometimes in paste, is traceable to another allied fungus. Dr. Murray describes this, and I quote his words. "A curious circumstance occurred in Padua, in August, 1819 ; the alimentary substance, called by the Italians *polenta*, is a compound of maize-flour, salt, and water. It was prepared in a family in the usual way, but was discovered to be full of red spots; this was cast away; but the red spots appeared in the next preparation which excited consternation and alarm. The blessing of the priest was implored, and given; still it was of no avail; prayers, fasts, and masses, were equally ineffectual. The gouttes of blood still appeared! The neighbours regarded the house and its inmates with fear and horror; in their opinion the polenta must have been made from some old corn, refused to the poor during the famine of 1819; and this was an evidence of the divine displeasure. A skilful botanist however, restored tranquillity and peace, by tracing the cause of alarm to a cryptogamic vegetation."

The cellar fungus or Mouse-skin is another member of this last class of fungi. This plant takes up its abodes in caves or cellars. A newly placed prop in a Derbyshire cavern was, in four years, draped with the felt formed by this mouse-skin (Raeodium cellare). It is in damp places that it is generally found.

Thus we come to an end of our selection of fungus species; having considered those most generally found and most easy of recognition. We have examined the distinctive features in the classes, the difference between simple and ascus contained spores marking the two great divisions; the exposed hymenium, characteristic of the first class (Hymenomycetes); the envelope marking the second class (Gasteromycetes), the prevalence of dust-like spores in the third class (Coniomycetes); and of the threads in the fourth (Hyphomycetes). Then the predominance of the spore bags in the fifth class (Ascomycetes); (the first class of the second, or Ascus division); and that of the threads in the sixth class (Physomycetes).

We have considered useful fungi—those good for food as the mushroom, chanterelle, morel and truffle; those useful in the preparation of food, as the yeast plant, the vinegar plant, blue cheese mould, etc.; those useful for other purposes, as the amadou of commerce, made from a polypore; snuff, made from another polypore; a trichogaster, used for dye; and the puff-balls, for preserving the lives of bees. We have looked with horror on poisonous agarics, on the dry rot, stinkhorn, moulds, bunts, smuts, mildews, and ergots, and seen how the worst of all is converted into a blessing by the blessing of God on man's ingenuity. Duly considering these things, I trust we shall not feel our time among the fungi wasted.

GENERAL INDEX.

	Page		Page		Page
A		Beech fern,	5	Chlorosperms,	162
Acrogens,	1	Black maidenhair,	23	Chondrus,	159
Adder's tongue,	41	Bladder fern,	31	Chorda,	135
Adiantum,	39	Bladder moss,	96	Chordaria,	139
Æcidium,	283	Blechnum,	35	Chroolepus,	176
Agaric,	244	Blight gooseberry,	287	Chylocladia,	149
Alaria,	134	—— potato,	289	Cinclidium,	95
Algæ,	128	Bloodstain,	160	Cinclidotus,	74
Alectoria,	229	Bog moss,	59	Cladonia,	232
Allosorus,	7	Boletus,	260	Cladophoræ,	168
Amanita,	244	Dorrera,	223	Cladostephus,	142
Amblyodon,	95	Bostrichia,	147	Clathrus,	276
Anacalypta,	70	Botrychium,	44	Clavaria,	266
Andræa,	58	Botrydium,	175	Climacium,	104
Anodus,	64	Botrytis,	289	Clitopilus,	251
Androgynum,	88	Brachyodus,	64	Clitocybe,	248
Anœctangium,	103	Brake,	37	Club fungus,	266
Antheridea of sea-		Bristle fern,	42	Club moss,	72
weeds,	129	Bristle moss,	64, 80	Codium,	165
Antitrichea,	104	Bryopsis,	167	Cœnothalami,	237
Apothecia of lichens,	188	Bryum,	88	Collar moss,	100
Apple moss,	97	Bulgaria,	300	Collema,	213
—— lurid,	99	Bunt,	285	Collybia,	248
—— naked,	99	Buxbaumia,	83	Conferva,	169
Arctoa,	65			Confervæ,	168
Arcyria,	278	**C**		Coniomycetes Class,	282
Aregma,	283	Calicium,	191	Cone fringe moss,	99
Arthonia,	192	Callithamnium,	162	Conostomium,	99
Ascomycetes class		Calocera,	268	Coprinus,	258
(fungi),		Calothrix,	171	Coral lichen,	232
Ascus of fungi,	242	Campilopus,	67	Coralline,	149
Ascus class,	249, 292	Candle lichen,	206	Cordiceps,	302
Aspergillus,	289	Candle snuff fungus,	305	Cord moss,	95
Asperococcus,	138	Cantharellus,	256	Cornicularia,	230
Aspidium,	16	Cap fungus,	242	Corticium,	266
Asplenium,	21	Carrageen moss,	159	Crab's claws,	149
Athalami,	237	Catenella,	160	Crnb's eye lichen,	203
Athyrium,	32	Catoscopium,	99	Craterellus,	265
Atrichum,	85	Cavern moss,	100	Crepidotus,	253
Auriculini,	265	Cells fungus,	242	Crostal,	207
		Ceramiaceæ,	161	Crotal,	207
B		Ceterach,	6	Cronania,	67
Bangia,	170	Cetraria,	221	Cronoria,	160
Bartramia,	97	Chætophora,	176	Crucibulum,	270
Batrachospermum,	179	Chantarelle,	256	Crustaceous lichens,	188
Beardless moss,	63, 75	Chara,	184	Cryptonemaceæ,	157

312   GENERAL INDEX.

D
		Page
Dacrymyces,	. .	271
Dædalia,	. .	263
Daltonia,	. .	106
Dasya,	. .	148
Delesseria,	. .	151
Desmarestia,	. .	136
Desmidiaceæ,	. .	174
Diatomaceæ,	. .	174
Dichelyma,	. .	107
Dicranodontium,	.	67
Dicranum,	. .	65
Dictyosiphon,	.	137
Dictyota,	. .	137
Didymium,	. .	278
Didymodon,	. .	71
Diphyscium,	. .	84
Disceliuin,	. .	99
Dissodon,	. .	100
Draparnaldia,	. .	176
Dry rot,	. .	203
Dudresnaia,	. .	160
Dulse,	. .	156
Dumontia,	. .	160
Dust class (fungi)	.	282

E
Earth moss,	. .	62
Earth star,	. .	276
Echinella,	. .	180
Ectocarpaceæ,	. .	142
Ectocarpus,	. .	143
Edible nests,	. .	158
Elachistea,	. .	140
Elvellacei,	. .	293
Encalypta,	. .	75
Endocarpon,	. .	195
Enteromorpha,	. .	169
Envelope class (fungi)		283
Epiphytes,	. .	283
Equisetum,	. .	48
Ergot,	. .	302
Evernia,	. .	226
Exidia,	. .	169
Extinguisher moss,	.	75

F
Fairy bath,	. .	298
Fairy fern,	. .	41
Fairy rings,	. .	257
Feather moss,	. .	111
Fern,	. .	2
—— allies,	. .	47
—— bladder,	. .	31
—— bristle,	.·	42
—— Hard,	. .	35
—— Heath,	. .	12
—— Holly,	. .	12
—— Lady,	. .	32
—— Prickly,	. .	10
—— Royal,	. .	42
Filmy fern,	. .	41
Fissidens,	. .	100
Fistulina,	. .	263
Five-leaved fork-moss,		71

		Page
Flat-leaved fork-moss,		100
Flowerless Plants,	.	1
Fontinalis,	. .	106
Fork moss,	. .	65
Fossil ferns,	. .	9
Four tooth moss,	.	83
Freshwater weeds,	.	174
Fringe moss,	. .	76
Frond moss,	. .	105
Frondose lichens,	.	188
Fucus,	. .	131
Funaria,	. .	96
Fungi,	. .	239
Fungus, pars of,	.	242
—— bird's nests,		278
—— candle snuff,		305
—— ear,	. .	270
—— hoof,	. .	307
—— jelly,	. .	269
Furcellaria,	. .	159

G
Gasteromycetes,	.	273
Gelidium,	. .	158
Geoglossum,	. .	296
Gigartina,	. .	158
Globe lichen,	. .	231
Gloiosiphonia,	. .	160
Gonidia of lichens,	.	187
Gooseberry blight,	.	287
Gracelaria,	. .	157
Grateloupia,	. .	158
Griffithsea,	. .	161
Grimmea,	. .	75
Gymnogougrus,	.	159
Gymnogramma,	.	6
Gymnostomum,	.	63
Gyrophora,	. .	219
Gulph weed,	. .	130

H
Hair moss,	. .	85
Hair-mouth moss,	.	70
Halidrys,	. .	130
Haliseris,	. .	138
Hard fern,	. .	35
Hart's tongue,	.	34
Hedwigia,	. .	75
Helvella,	. .	294
Himanthalia,	. .	133
Hirneola,	. .	270
Hoof fungus,	. .	307
Horse tail,	. .	48
Hydnum,	. .	264
Hygrophorus,	. .	255
Hymenium of fungus,		242
Hymenocbæte,	.	266
Hymenomycetes,	.	242
Hymenophyllum,	.	43
Hypholoma,	. .	254
Hyphomycetes class,		288
Hypnea,	. .	157
Hypnum,	. .	111
Hypocrea,	. .	*305
Hypogei,	. .	273

		Page
Hypoxilon,	. .	305
Hyssop on the wall,		63
Hysterium,	. .	302

I
Idiothalami,	. .	237
Iodine,	. .	133
Iridæa,	. .	160
Isariacei,	. .	288
Isidium,	. .	23
Isothecium,	. .	105
Isoetes,	. .	51
Ivory lichen,	. .	227

J
Jania,	. .	150
Jelly fungus,	. .	269
Jersy fern,	. .	6
Jew's ear,	. .	270

K
Kallymenia,	. .	160
Kelp,	. .	131

L
Lady-fern,	. .	32
Lamellæ of fungus,		242
Laminaria,	. .	134
Lastræa,	. .	16
Laurencia,	. .	148
Leathesia,	. .	139
Lecanora,	. .	202
Lecidea,	. .	200
Lecythea,	. .	284
Lentinus,	. .	258
Lenzites,	. .	258
Leotia,	. .	295
Lepiota,	. .	244
Lepraria,	. .	200
Leptobryum,	. .	88
Leskea,	. .	105
Leucobryum,	. .	295
Leucodon,	. .	103
Lichina,	. .	237
Lichens,	. .	187
Lichen, candle,	.	206
—— dog,	. .	215
—— goblet,	. .	191
—— jelly,	. .	213
—— map,	. .	200
—— scale,	. .	205
—— scurf,	. .	205
—— socket,	. .	217
—— spangle,	. .	197
—— wart,	. .	194
—— writing,	. .	193
Lobeless plants,	.	1
Lungs of the oak,	.	209
Lycoperdon,	. .	276
Lycopodium,	. .	52
Lyngbye s weed,	.	171

M
Maiden-hair,		40

GENERAL INDEX. 313

	PAGE		PAGE		PAGE
Male fern,	81	Orchil,	222	Scolopendrium,	34
Meesia,	95	Orthodontium,	88	Screw moss,	72
Melanosperms,	129	Orthotrichum,	80	Scyphophorus,	233
Melobesia,	150	Oscillatoria,	171	Scytonema,	182
Merulins,	263	Osmunda,	42	Seaweeds,	128
Mildew,	306	**P**		—— parts of,	129
Mesogloia,	139	Padina,	138	Sea endive,	138
Mitre fungus,	294	Palmella,	183	—— oak,	129
Mitrula,	294	Parsley fern,	7	—— thong,	133
Mnium,	92	Peacock weed,	138	—— whipcord,	135
Moonwort,	44	Perisporiacei,	306	Shield fern,	16
Morchella,	293	Peziza,	297	Siller cup,	279
Morel,	293	Pill wort,	51	Smut,	285
Mosses,	57	Phascum,	62	Socket lichen,	217
Moss, beardless,	63, 75	Pholiota,	251	Solorina,	217
—— bog,	57	Phyllophora,	159	Snow, red,	183
—— bristle,	64, 80	Physcomitrium,	96	Spangle lichen,	197
—— cone,	99	Physomycetes class,	302	Spathularia,	295
—— cord,	95	Pleurotus,	250	Spawn fungus,	242
—— extinguisher,	75	Plocamium,	152	Sphacelaria,	143
—— earth,	62	Pogonatum,	85	Sphagnum,	59
—— five-leaved,	71	Polyides,	159	Sphærococcus,	157
—— fork,	65	Polypody,	3	Sphærophoron,	231
—— four-toothed,	83	Polyporus,	261	Sphæria,	302
—— fringe,	76	Polystichum,	10	Spiloma,	196
—— hair,	85	Polysiphonia,	147	Splachnum,	100
—— hairmouth,	70	Polytrichum,	86	Spleenwort,	21
—— screw,	74	Potato blight,	289	Spores, fungus,	242
—— swan neck,	67	Pottia,	70	Sporidia,	242
—— twin tooth,	71	Preserve mould,	307	Squamaria,	205
—— yoke,	82	Prickly fern,	10	Stag's horn moss,	52
Moulds,	288	Protococcus,	183	Star slough,	183
Mould cheese,	289	Psalliota,	253	Stereum,	265
—— dung,	289	Pteris,	37	Sticta,	209
—— herb,	289	Ptilota,	161	Stinkhorn,	274
—— bread,	307	Puccinea,	283	Swan neck moss,	67
Mouse-skin,	303	Puff ball,	276	Sweet dulse,	160
Mucedines,	307	Punctaria,	138	Sweet tangle,	134
Mungo Park's moss,	100	Pycnophycus,	131	Swine tang,	133
Mushroom,	253				
Mycena,	250	**R**		**T**	
Myrionema,	140	Racomitrium,	76	Tangle,	128
Myriotrichia,	144	Ralfsia,	139	—— sweet,	134
Myxogastres,	278	Ramalina,	227	Taoni,	138
		Red snow,	183	Tayloria,	100
N		Reindeer lichen,	232	Tetradontium,	83
Naccaria,	160	Reticularia,	278	Tetraphis,	83
Naucoria,	251	Rhodosperms,	143	Tetraplodon,	100
Neckera,	107	Rhodymenia,	156	Theleholus,	286
Nectria,	306	Rhytisma,	302	Thelephora,	265
Nidulariacei,	278	Rivularia, freshwater,	180	Thread class (fungi),	288
Nitophyllum,	151	—— marine,	171	Thread moss,	88
Nostoc,	183	Rocella,	222	Thyme thread moss,	92
Nullipores,	150	Rock hair,	229	Timmia,	87
Nyctalis,	256	Royal fern,	42	Tortula,	72
		Rytiploca,	147	Torula,	283
O				Trametes,	262
Oak fern,	5	**S**		Tree moss,	105
Odonthalia,	156	Sap balls,	26	Tremella,	269
Œdipodium,	100	Sargassum,	130	Tricholoma,	247
Old man's hair,	229	Scale lichen,	205	Trichomanes,	42
Omphalia,	250	Schistidium,	76	Trichostomum,	70
Onygena,	307	Schistostega,	100	Trichia,	278
Opegrapha,	193	Scleroderma,	277	Truffle,	300
Ophioglossum,	41				

	PAGE		PAGE		PAGE
Truffle bath,	274	**V**		Weissia,	63
Tuber,	300	Variolaria,	96	Wing moss,	103
Twin-tooth moss,	71	Vaucheria, freshwater,	175	Woodwitch,	274
		—— marine,	167	Wrapper (fungus),	243
U		Veil of fungus,	242		
Ulva,	170	Velum,	242	**X**	
Ulvaceæ,	169	Verpa,	295	Xylaria,	305
Ulva, freshwater,	178	Verrucaria,	194		
Umbilicaria,	220	Vesicles of seaweeds,	129	**Y**	
Urceolaria,	197	Vinegar plant,	290	Yeast plant,	286
Uredo,	284	Volva,	234	Yoke moss,	82
Urn-moss,	57	Volvaria,	251		
Usnea,	227			**Z**	
Ustilago,	285	**W**		Zasmidium,	
		Wall rue,	26	Zygnema,	179
		Water screw-moss,	74	Zygodon,	83

INDEX TO PLATES.

PLATE I. (*Facing page* 1.)

1. Common Polypody. 2. Oak Polypody. 3. Beech Polypody. 4. Limestone Polypody. 5. Scaly Spleenwort. 6. Gymnogramma. 7. Angular Prickly Shield fern. 8. Parsley fern.

PLATE II. (*Facing page* 15.)

1. Male fern. 2. Spreading Shield-fern. 3. Spiny Shield-fern. 4. Heath Shield-fern. 5. Marsh Shield-fern. 6. Crested Shield-fern.

PLATE III. (*Facing page* 21.)

1. Alternate-leaved Spleenwort. 2. Forked S. 3. Black Maiden-hair. 4. Sea S. 5. Black-stalked S. 6. Wall rue. 7. Bladder-fern. 8. Lady fern. 9. Hart's tongue. 10. Hard fern.

PLATE IV. (*Facing page* 37.)

1. Brake. 2. Woodsia. 3. Maiden-hair. 4. Tunbridge Filmy-fern. 5. Bristle fern. 6. Royal fern. 7. Moonwort. 8. Adder's tongue. 9. Jersey Adder's tongue.

PLATE V. (*Facing page* 57.)

2. Curved-leaved Beardless moss. 3. Blindia. 4. Broom Fork Moss. 5. Silky F. 6. Leafy Buxbaumia. 7. Purple Fork moss. 8. Fine-leaved Distichium. 9. Curve-leaved Hair Mouth moss. 10. Bent-leaved Didymodon. 11. Water Screw-moss. 12. Fringed Hedwigia. 13. Rock Andræa. 14. Grey-cushion Grimmia. 15. Oval-fruited G. 16. Hoary Fringe moss. 17. Slender mountain F. 18. Dark Mountain F.

PLATE VI. (*Facing page* 69.)

1. Green tufted Weissia. 2. Bent leaved W. 3. Don's Bristle moss. 4. Common Pottia. 5. Wall Screw-moss. 6. Awl leaved do. 7. Fallacious do. 8. Twisted do. 9. Great Hairy do. 10. Müller's do. 11. Pellucid Four tooth moss. 12. Budheaded Thread-moss. 13. Golden Thread moss. 14. Haller's Apple moss. 15. Curve-stalked do. 16. Lurid Apple-moss. 17. Cavern moss.

PLATE VII. (*Facing page* 103.)

1. Tetraplodon. 2. Squirrel-tail Leucodon. 3. Tall Anomodon. 4. Fern-like Feather moss. 5. Flat-leaved Neckera. 6. Greater Water moss. 7. Alpinedo. 8. Beaked Water Feather moss. 9. Neat do. 10. Neat Mountain do.

PLATE VIII. (*Facing page* 106.)

1. Silky Leskea. 2. Foxtail Frond-moss. 3. Tamarisk Feather moss. 4. Striated F. 5. Triangular F. 6. Waved F. 7. River F. 8. Wood F. 9. Shining Hookeria. 10. Creeping Feather moss.

PLATE IX. (*Facing page* 127.)

1. Sea Oak. 2. Common Cystoseira. 3. Serrated Fucus. 4. Sea Whipcord. 5. Edible Alaria. 6. Fingered Laminaria. 7. Thorny Desmarestia. 8. Gulph weed. 9. Sea-thong.

PLATE X. (*Facing page* 137.)

1. Sweet Laminaria. 2. Peacock weed. 3. Forked Dictyota. 4. Tube Leathesia. 5. Funnel Dictyosiphon. 6. Whip-like

INDEX TO PLATES.

Chordaria. 7. Broom-like Sphacelaria. 8. Littoral Ectocarpus. 9. Feathery Sphacelaria. 10. Verticled Cladostephus. 11. Fucus Elachistea. 12. Compressed Asperococcus.

PLATE XI. (*Facing page 145.*)

1. Toothed Odonthalia. 2. Elongated Polysiphonia. 3. Dark do. 4. Bonnemaison's weed. 5. Articulated Chylocladia. 6. Red Jania. 7. Bloody Delesseria. 8. Oak do. 9. Plocamium.

PLATE XII. (*Facing page 155.*)

1. Dulse. 2. Pimpled Gigartina. 3. Carrageen. 4. Feathery Callithamnion. 5. Bundle Furcellaria. 6. Edible Iridea. 7. Feathery Ptilota. 8. Knotted Ceramium. 9. Red do. 10. Shrubby Callithamnion.

PLATE XIII. (*Facing page 187.*)

1. Rose Bœomyces. 2. Red do. 3. Short-stalked Goblet lichen. 4. Black G. 5. Golden G. 6. Greek Opegrapha. 7. Black O. 8. Variable O. 9. Submerged Endocarpon. 10. Grey E. 11. Common Pertusaria. 12. Brain Opegrapha. 13. Elegant O. 14. Rock O. 15. Birch O. 16. Starry O. 17. Speckled O. 18. Rusty Goblet-lichen.

PLATE XIV. (*Facing page 199.*)

1. Rock Lecidea. 2. Map L. 3. Sunken L. 4. Black L. 5. Yellow L. 6. Green L. 7. Bog L. 8. Limestone Urceolaria. 9. Oak Lungs. 10. Crab's eye. 11. Cudbear. 12. Pitted Sticta. 13. Brown Scurf-lichen. 14. Black S. 15. Wall Squamaria. 16. Stone S. 17. Sulphur Parmelia. 18. Crotal. 19. Rock P. 20. Burnt P.

PLATE XV. (*Facing page 212.*)

1. Great Collema. 2. Crisp C. 3. Green Socket-lichen. 4. Many leaved Gyrophora. 5. Burnt G. 6. Fleecy G. 7. Iceland moss. 8. Snow Cetraria. 9. Glaucous C. 10. Fucus-like Roccella. 11. Hairy Borrera. 12. Dwarf B. 13. Branny B. 14. Dog lichen. 15. Wall Parmelia.

PLATE XVI. (*Facing page 225.*)

1. Evernia. 2. Bundle Ramalina. 3. Ash R. 4. Rock R. 5. Old Man's hair. 6. Wooly-horn lichen. 7. Brittle Globelichen. 8. Reindeer moss. 9. Branched Stereocaulon. 10. Common Cup lichen. 11. Scaly C. 12. Coral C. 13. Finger C. 14. Elegant C. 15. Compressed Globe lichen. 16. Hairy Usnea.

PLATE XVII. (*Facing page 237.*)

1. Fly Agaric. 2. Purple A. 3. Hollow-stemmed Collybia. 4. Oak-leaf C. 5. Rose Mycena. 7. Cup Omphalia. 8. Dingy Pleurotus. 9. Golden Pholiota. 10. Variable Crepidotus. 11. Verdigris Psalliotta. 12. Olive-gilled Hypholoma. 13. Inky Coprinus. 14. Starry C. 15. Red Russula. 16. Orange Omphalia. 17. Pale Tricholoma. 18. Edible Mushroom. 19. Cornucopia Cratellus.

PLATE XVIII. (*Facing page 260.*)

1. Yellow Sap-ball. 2. Lurid do. 3. Moss Cyphel. 4. Scaly Polypore. 5. Fir P. 6. Corky Merulius. 7. Fistulina. 8. Spread Hydnum. 9. Purple Corticium. 10. Brown Hymenochæte. 11. Yellow Clavaria. 12. Amethyst C. 13. Furrowed C. 14. Candle C. 15. Golden Calocera. 16. Bloody Stereum. 17. Glandular Exidia. 18. Jews' Ear. 19. Orange Tremella.

PLATE XIX. (*Facing page 273.*)

1. Bath Truffle. 2. Stinkhorn. 3. Hairy Earth-star. 4. Common Puff-ball. 5. Pear P. 6. Larger P. 7. Red Arcyria. 8. Common Trichia. 9. Striped Bird's Nest. 10. Bell B. 11. Crucible B. 12. Edible Chanterelle. 13. Edible Morel. 14. Mitre-like Helvella. 15. Marsh Mitrula. 16. Shining Leotia. 17. Common Spathularia. 18. Olive Geoglossum. 19. Shining G. 20. Soldier Cordiceps.

PLATE XX. (*Facing page 292.*)

1. Scarlet Peziza. 2. Orange P. 3. Hotbed. P. 4. Green P. 5. Veined P. 6. Claret P. 7. Yellow P. 8. Black and Rusty P. 9. Verdigris P. 10. Hairy P. 11. Ground P. 12. Bulgaria. 13. Shapeless Hypoxilon. 14. Candle-snuff Xylaria. 15. Clumsy Xylaria.

INDEX TO WOODCUTS.

A

	PAGE
Adiantum,	37
Agaric, spore of, and gill,	242
Allosorus,	7
Andræa, fruit and leaf,	59
Apothecia of lichens,	195
Apothecia of Nephroma,	215
Ascus of Peziza,	242
Asplenium,	21
Atrichum Undulatum,	79

B

	PAGE
Bartramia,	91 & 96
Bartramidula,	91
Batrachospermum,	178
Blechnum,	35
Blight, gooseberry	28
Blight, wheat	286
Blue cheese mould,	289
Bog mosses,	60
Boletus, hymenium of,	264
Botrichium,	44
Botrydium,	178
Branch of globe lichen,	215
Bryum,	79
Bryum magnified	96
Bunt,	286

C

	PAGE
Calothrix,	164
Ceramium,	162
Ceterach,	7
Chanterelle,	242
Cheese mould,	289
Cladophora,	164
Codium,	164
Conferva,	164 & 175
Cystopteris,	30
Cystoseira,	131

D

	PAGE
Dicranum,	73
Dog lichen,	215
Dung mould,	289

E

	PAGE
Equisetum,	46

	PAGE
Encalypta,	73
Endocarpon, section of,	195
Enteromorphæ,	164

F

	PAGE
Fissidens,	91
Fistulina,	264
Fucus,	131
Funaria,	91 & 96

G

	PAGE
Gill of Agaric,	242
Globe lichen, branch of,	215
Gonidium of lichen,	195
Gooseberry blight,	286
Grimmia,	86
Gymnostomum,	63

H

	PAGE
Halidrys,	126
Hedwigia,	86
Herb mould,	289
Hydnum,	264
Hymeniums, various	242, 262
Hymenophyllum,	40

I

	PAGE
Isoetes,	46

J

	PAGE
Jungermannia,	117

L

	PAGE
Lastrea,	15
Lichen, dog, etc.,	215
Lichen, spore, etc.,	195
Liverworts,	117
Lycopodium,	46

M

	PAGE
Marchantia,	117
Mnium,	91
Mould, cheese,	289
Mould, dung,	289
Mould, herb,	289

O

	PAGE
Opegrapha,	195

	PAGE
Ophioglossum,	44
Orthotrichum,	79, 86
Osmunda,	42

P

	PAGE
Paludella,	94
Pertusaria,	195
Peziza,	242
Phascum,	63
Physcomitrium,	70, 94
Plocamium,	152
Polypodium,	3
Polyporus,	264
Polystichum,	12
Polytrichum,	79, 83
Porphyra,	164
Pottia,	70
Protococcus,	178
Pteris,	37

R

	PAGE
Rivularia,	164

S

	PAGE
Sargassum,	129
Scolopendrium,	34
Sphagnum,	59, 60
Splachnum,	91

T

	PAGE
Tetraphis,	70
Tortula,	73
Trichomanes,	42

U

	PAGE
Urceolaria,	195

V

	PAGE
Vaucheria,	175
Verrucaria,	195

W

	PAGE
Weissia,	76
Wheat blight,	286
Woodsia,	13

Z

	PAGE
Zygodon,	79

www.ingramcontent.com/pod-product-compliance
Lightning Source LLC
Chambersburg PA
CBHW020228240426

43672CB00006B/454